Reconciliation of Geometry and Perception in Radiation Physics

FOCUS SERIES

Series Editor Piotr Breitkopf

Reconciliation of Geometry and Perception in Radiation Physics

Benoit Beckers
Pierre Beckers

iSTE

WILEY

First published 2014 in Great Britain and the United States by ISTE Ltd and John Wiley & Sons, Inc.

ISTE Ltd
27-37 St George's Road
London SW19 4EU
UK

www.iste.co.uk

John Wiley & Sons, Inc.
111 River Street
Hoboken, NJ 07030
USA

www.wiley.com

Library of Congress Control Number: 2014939768

British Library Cataloguing-in-Publication Data
A CIP record for this book is available from the British Library
ISSN 2051-2481 (Print)
ISSN 2051-249X (Online)
ISBN 978-1-84821-583-2

Printed and bound in Great Britain by CPI Group (UK) Ltd., Croydon, Surrey CR0 4YY

Contents

Introduction

If we must now reconcile geometry and perception, it is because a separation has occurred. As its etymology suggests, geometry was born in Greece about 2,500 years ago. From the figures (in particular, straight lines and conical curves) and relationships (especially parallelism and orthogonality) that structure human vision, the concern was to describe an essentially visible world. As we recall in Chapter 1 of this book, this visual geometry, first in competition with an essentially auditory arithmetic, quickly articulated with it, then took over, to the point that we have witnessed a geometrization of mechanics, of color (Newton) and, finally, of the world itself (Galileo and Kepler).

Then, in the early 19th Century, physics and mathematics turned away from this visual world. This would be, afterward, a problem for those precisely interested in the sensory world, either for itself, or to represent the results of these now so abstract sciences and mathematics. Should it be necessary to specify a date to symbolize the passage to abstraction, a particularly significant candidate would be that of 21 March 1816, when a royal decree of the recently restored French monarchy dissolved the Academy of Sciences to reinstate it without any members involved with the Revolution and the Empire, including Gàspard Monge.

With his descriptive geometry, Monge created a tool able to represent, for the first time, the infinite objects of Euclid (straight lines and planes) so that he could easily solve on the sketch plan the three-dimensional problems that are difficult to represent in space. He also gave a correct explanation of colored shadows. Indeed, he proposed a visual and mechanical interpretation of the world that the Greeks would not have discarded.

Monge was also the first director of the Ecole Polytechnique. Among his students, Jean-Victor Poncelet participated in the Russian campaign of 1812. Taken as prisoner, he spent 2 years in the prison of Saratov on the Volga. This is where he imagined the foundations of projective geometry.

Still on the Volga River, 500 km upstream, Nikolai Ivanovich Lobachevsky was then completing his studies at the University of Kazan and developing an "imaginary geometry", which was the first of the so-called "non-Euclidean" geometries, as Carl Friedrich Gauss would qualify them later. For Lobachevsky, Euclidean geometry was a limiting case where, as he wrote in 1829, "light rays serve as straight lines; it is not a question of logic, but of physics, which cannot be answered experimentally. As our measurements show no change from Euclidean geometry, we can assume that this one governs our space" [FER 11]. To check if "penetrating extremely remote areas of the universe, our measuring devices would detect such variations", Lobachevsky measured the interior angles of the triangle formed by the Earth, the Sun and the star Sirius. With the experimental means of the time, he was unable to verify whether their sum was equal to 180° (Euclidean space) or lower (hyperbolic space), but his approach clearly anticipated the sensory world being surpassed, realized some decades later by the combination of experimental physics and abstract mathematics.

Let us go back to Paris and the year 1816. Monge's successor at the Academy of Sciences is Augustin Louis Cauchy, trained, as Poncelet, at the Ecole Polytechnique. Everything separates Monge and Cauchy. The first, born in 1748, spent most of his career under the Old Regime before joining the Revolution, then Bonaparte, who he followed in Egypt and to which he remained faithful until the end. Cauchy, son of the Revolution (he was born in August 1789), formed in republican schools, but from a Catholic family, would be a royalist, a legitimist, and voluntarily went into exile in 1830 after the fall of Charles X.

His works on holomorphic functions are often considered as the true break with visual geometry. During the 1820s, Cauchy in France and Gauss in Germany ruled on mathematics. In particular, they developed the study of the complex plane.

The projective plane, the complex plane, as well as spaces with a constant negative curvature (hyperbolic geometry of Lobachevsky) or a constant positive one (elliptic geometry of Bernhard Riemann, a disciple of Gauss) were the new playgrounds for geometries often competing, and moving decidedly away from perception.

At the end of the century, two great syntheses ordered these ideas. In 1872, in his "Erlangen Program", Felix Klein described all geometry as the search for the invariants of some transformation group. In 1899, in his "Foundations of Geometry" David Hilbert axiomatically restructured Euclid's elements.

Geometry thus seems totally abstracted from perception, looking for new physics also abstracted from the sensory world, which would occur only in the 20th Century.

In the 1800, William Herschel found that decomposed sunlight through a prism heats outside the spectrum of visible light, beyond the red. The following year, Johann Wilhelm Ritter discovered ultraviolet. These invisible waves, capable of heating more than visible light (infrared) or acting on the silver chloride (ultraviolet), are phenomena that we cannot see, but that can be measured, such as gravity, magnetism and electricity.

The physics of the 19th Century was first in the continuity of the previous two centuries, with an ever greater effort brought to measurement and algebraic formulation. In 1811, Joseph Fourier equated heat. In the second part of the century, James Clerk Maxwell unified electromagnetism. Throughout this period, evidence accumulated to the detriment of the universality of Newtonian mechanics, until the work of Max Planck on black body and quanta, presented in 1900.

These great advances in physics created a great victim: geometrization, that is to say the setting in space. Let us take the example of acoustics. The 18th century is the century of the treatise on the opera house, looking for its ideal form. Should the public be disposed in a circle, an ellipse or a horseshoe? What is the ideal size of the room? What form should the tiers, the balconies and the ceiling adopt? At the time, the sound field was not quantified, and reasoning was purely qualitative. What could not be determined for the ear was for the view: the integrator tool was still the central perspective, as in the spectacular views of Claude Nicolas Ledoux and Étienne-Louis Boullée.

However, the 19th Century was unable to apply its new equations to such complex geometric shapes. For natural light, it was even worse. Buildings of the 18th Century were expertly open to daylight so that employees could work well. In those of the next century, due to advances in lighting, openings on the sky were virtually lacking. Everywhere, industrialization was requiring simpler, more manageable, forms and progress in equipment poorly compensated the lack of shape design.

The study of perception was not neglected, quite the contrary. Many studies were performed on color (including those of Maxwell), and Hermann von Helmholtz transfigured musical acoustics (On the Sensation of Tones, 1863) with an experimental resonators device allowing him to explain the formation of the musical scale [MEU 10].

Such research found its framework in a recent discipline, physiology, soon complemented by experimental psychology. However, these disciplines, increasingly numerous and increasingly compartmentalized (musicology, physical and physiological acoustics, psychoacoustics, etc.), were less oriented to applications. Helmholtz stated that "Whoever in the pursuit of science seeks after immediate practical utility may rest assured that he seeks in vain" (Academic Discourse, Heidelberg, 1862).

In the 20th Century, some scientists developed new concepts and tools to control the energy fields in real-world projects: invention of heliodons and solar diagrams, acoustic ray tracing, etc. However, such activities were considered to be totally marginal from the viewpoint of both mathematics and physics, as they were distant from the new developments achieved far away from the human scale.

In the last third of the century, finally, the development of electronics, computer graphics and numerical methods upset these partitioned disciplines. The main change is to visualize what is not visible, starting with infrared (thermography). Like in the Renaissance, all recovers its place in space and time.

For us, geometric projection and physical radiation are both faces of the same problem. Radiation is instantaneous (light), delayed (acoustic) or inertial (thermal). It takes place in a scene composed mainly of surfaces, which are provided with some material properties (e.g. reflection coefficients).

If we observe now, beyond the 19th Century, the history of representation, we can see an early split into two main ways. The first consists of unfolding which can be (development), cutting in volumes and, if projection is necessary, maintaining at all costs the proportions and parallels (axonometric perspective), even multiplying the projection planes to deform the objects of the scene (technical drawing) as little as possible. This trend began with prehistoric parietal paintings up until modern tomography, through Egyptian drawings, Persian or Chinese parallel perspectives, to the method of Monge. It continues to be productive today (level sets), but this is not what we discuss here.

The second way, already evident in the optics of Euclid and in Renaissance perspective works and theories, is to accept the inherent three-dimensionality of our space, as we live, hear and see it. This is the way of sensory geometry, finally reconciled with the perceptions that gave it birth, the sound and light whose serene waves radiate in it.

We will start with music.

1

Discovering the Central Perspective

1.1. The musical scale

The frequency aspect of the system of musical composition is based on a physical property; the harmonic structure of periodic sounds; an arithmetic property; almost perfect equality between the 12th term of the triple progression and the 19th power of 2; and a simple experimental device, for example, a string fixed at both ends and stretched on a graduated scale, so that it can be divided accurately, pressing on one of its graduations.

Our ears easily verify some properties: the shorter the string, the sharper its sound; the simple divisions (half, two-thirds, three-quarters, etc.) produce, with respect to the entire string, consonant intervals (respectively, the octave, the fifth, the fourth, etc.); if the same division is repeated on the shortened string, then the same interval is perceived (for example, if the string is divided into two, then into four, then into eight, three ascending octaves are successively heard).

With a very long and heavy string, whose vibration is very slow, it can be inspected visually that the perceived height is proportional to the vibration frequency and, therefore, inversely proportional to the string length. So, what the ears identify as the interval between two pitches corresponds physically to a ratio, i.e. a multiplicative factor, whose product by the frequency of the lower pitch of the interval gives the frequency of its top pitch.

With a string of length L, we can produce the following simple intervals between the entire string and its division:

Division	Ratio	Interval
L	1	unison
L/2	2/1	octave
2L/3	3/2	fifth
3L/4	4/3	fourth
4L/5	5/4	major third
5L/6	6/5	minor third
8L/9	9/8	tone

The ratio series is interrupted before including the number 7. The following intervals would be too close together, so a jump is done directly to the tone, which is already dissonant: for two singers, it is extremely difficult to sing the same melody at a distance of one tone, and the result is heard as a "friction", i.e. dissonant. In contrast, the thirds are consonant, like further higher intervals, until the octave, the most consonant of all. When a man and a woman try to sing together in unison, an octave separates their voices. There is no doubt that the octave has always been perceived as an identifier interval: doubling the frequency of a sound only modifies its absolute pitch (that is, a jump of octave), not its relative one within a set of sounds.

Therefore, the problem of the musical scale is how to fill the octave with a maximum of consonant intervals in a logical way that will afterward allow the largest number of operations. Whenever a ratio takes us beyond the limits of the octave, we can divide by two, without changing the nature of the corresponding interval.

We find, for example, that the sum of a fifth and a fourth gives an octave because the product of their ratios gives the octave one: $3/2 \times 4/3 = 2$. The octave is also decomposed into a fourth, a major third and a minor third ($4/3 \times 5/4 \times 6/5 = 2$), or two fourths and a tone ($4/3 \times 4/3 \times 9/8 = 2$). More curious is the result of dividing the string into six equal parts (ratios 1, 6/5, 6/4, 6/3, 6/2 and 6/1), which produces the minor chord intervals (minor third, fifth and octave).

However, these early observations do not allow us even to boot up the composition of a full scale. To do this, we need a generalizable procedure. Around the 6th Century BC, the Pythagorean School moved to Europe a reasoning already well known in the East, possibly in Egypt, definitely in Chaldea and earlier in China.

The idea is to use only the two more consonant intervals: the octave and the fifth. Starting at any pitch (frequency f), which defines the octave to fill (f to 2f), we ascend by the interval of a fifth to a second pitch (3/2 f). A second fifth is then applied, giving a third pitch (9/4 f). As this is beyond the octave to fill, we lower it by an octave (9/8 f). And so on, progressing by fifths and regressing by octaves, we find the 12th pitch. This is almost equal to the first one due to the following arithmetic property:

$$3^{12} = 531441 \approx 524288 = 2^{19}$$

In other words, 12 fifths are almost equal to 7 octaves:

$$(1.5)^{12} = 129.746338 \approx 128 = 2^{7}$$

The ratio corresponding to the small difference is called a Pythagorean comma, and it is 312/219 = 1.013643. Without this, the problem would be solved: once ordered within the octave, the 12 notes are all separated by the same interval, the semitone (halftone). But it is not, and one of the semitones is distinguished from the others by a Pythagorean comma. All intervals that include it differ equally from the same intervals built without including it. Still, the scale formed by progression of fifths is excellent because the 12 fifths that can be formed in it are accurate, and the fifth is a very important interval for music, the most important after the octave.

However, the approximation of the comma could not satisfy the Chinese, who considered the *lu*, i.e. the ordered sound, as a principle of universal equilibrium [LIE 01]. So, they tried to reduce the comma, bringing further progression of fifths. In the 1st Century BC, King-Fang showed that 53 fifths are almost equal to 31 octaves, thus reducing the comma to a value of 1.002, almost unitary. But a scale of 53 notes is unmanageable. The Arabs, lovers of arithmetic, implemented lengthy scales on their most sophisticated instrument, the Qanun. But most instruments do not allow so much subtlety and less so the human voice.

After 12 fifths, we checked in Figure 1.1. that the difference is 0.0136 (corresponding to the Pythagorean comma); it is slightly reduced after 41 fifths, and even more so after 53, as discovered by King-Fang. Khien-Lon-Ki, the 5th Century AD astronomer, discovered the similar cases of 306 and 359 fifths. Due to the computer, we can now propose the prodigious scale of 665 notes, almost perfect, but perfectly useless.

Still in China, another system was then conceived. Since the scale must have 12 notes, why not give the value of the 12th root of 2 directly to the halftone? That is already the idea of the *equal temperament*, where the comma is equally shared among all intervals. In the 18th Century, when the Jesuit Joseph-Marie Amiot visited the empire, the Chinese were able to evaluate the terms of the geometric progression of ratio 12√2 with an accuracy of four decimals and to apply it perfectly to bamboo tubes. With the *yellow bell*, whose dimensions and composition were perfectly defined, they also possessed a universal diapason.

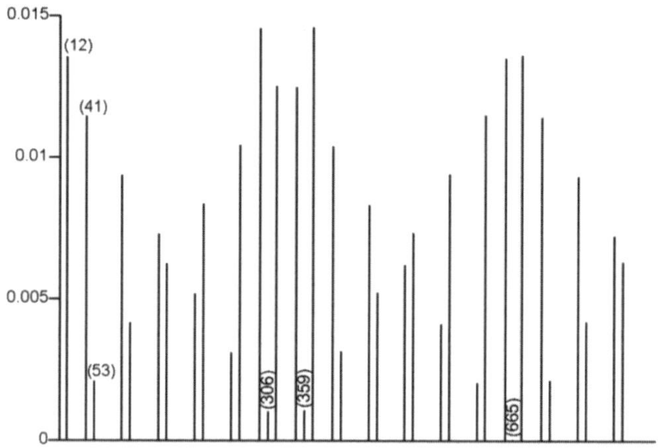

Figure 1.1. *King-Fang experience*

The problem with equal temperament is that all intervals are – very slightly – false. The more serious problem with all of the theories just discussed is that they are ultimately more arithmetic than musical: Plato already reproached the Pythagoreans for their excessive mathematization of music. In the mid-4th Century BC, Aristoxenus of Tarentum, using the same reproach, founded his own definition of the musical scale on another principle, more physical: the harmonic structure of sound.

So far, we have assumed two elementary mathematical entities: pure sounds (having a single frequency) and simple intervals. The problem is that their physical realization is not directly present in nature.

Most natural sounds – thunder, waterfall, etc. – have such a rich and disordered frequency structure that our ears cannot even characterize a dominant pitch. However, few sounds do have pitch: the chirping of birds, the wind through the reeds, etc., and the most important fact is that human industry can easily imitate these sounds and improve them.

The more refined sounds of music are not pure: they always cover a wide range of frequencies. However, they are periodic, i.e. they repeat their pattern in regular intervals. Fourier theory shows that any periodic waveform is decomposed into a series of harmonics that are integer multiples of the fundamental frequency. In a periodic sound, the fundamental frequency gives the root note, while the relative intensities of harmonics produce the timbre. All these frequencies are merged into a single sound, with its particular timbre, but, unlike what happens with mixing colors, mixed frequencies are perceived individually. Two physical phenomena reinforce this perception: resonance and beating.

If we produce a strong, short sound beside a harp, each string of the harp whose natural frequency is equal to any frequency emitted by the sound comes into *sympathetic resonance*. The vibrations of the harp inform us about the harmonic composition of the sound.

When two sounds are close in pitch but not identical, the difference in frequency generates a *beating*: the intensity varies, like in a tremolo, as the sounds alternatively interfere constructively and destructively. This phenomenon is used to tune instruments. In a fifth out of tune, for example, the second harmonic of the lower note will produce a beating with the top note. Setting the fifth, this beating will slow down until it disappears.

The physical properties of harmonics, resonance and beating together explain why the human ear appreciates simple intervals and why it is extremely sensitive to imperfections. But something happens.

Here is the list of the first harmonics of a periodic sound of any frequency f, with the intervals between each harmonic and the fundamental:

Harmonic	Ratio	Interval
f	1	fundamental
2f	2	octave
3f	3/2	fifth
4f	2	octave
5f	5/4	major third
6f	3/2	fifth
7f	7/4	~ diminished/minor seventh

First observation: the first six harmonics, which are consonant, form a major chord (major third, fifth and octave). When multiple instruments play this chord, they reproduce the harmonic structure of periodic sound. It is also when the triple series of harmonics best overlaps, hardly generating friction. Among all the chords of three different sounds, this is the most consonant.

Second observation: the order of the harmonics is not exactly the same as that of the simple intervals previously established: here, the major third (ratio 5/4) appears long before the fourth (ratio 4/3). In medieval polyphony, chords of fourths were considered to be very consonant. In the Renaissance, they disappeared completely, being substituted by chords of fifths and thirds. With little exaggeration, we can say that with this mutation, music went from being essentially arithmetic to essentially harmonic.

Third observation: Aristoxenus theory, based on the harmonic structure of sound, explains much better than that of the Pythagoreans, based on simple intervals, why certain intervals naturally sound consonant to the ears.

The double generation of the scale by arithmetic and physics properties is not sufficient to give rise to a composition system based on the consonant/dissonant contrast. On the one hand, the harmonic structure explains consonances, but remains static, because it cannot raise a logical development of complementary dissonances. Moreover, the progression by fifths offers a full 12 semitones scale, but this scale is also very static because it remains identical whatever the note from which we start to play it.

To make the harmony dynamic, it is necessary to reduce the scale in order to deploy two types of intervals instead of the single semitone. This is achieved by limiting the progression of fifths to five notes (e.g. F – C – G – D – A) or seven notes (e.g. F – C – G – D – A – E – H). Once the notes within the octave are rearranged, we obtain, in the first case, the pentatonic scale, which consists of only two tones and minor thirds, and in the second case, the heptatonic scale, which consists only of tones and two semitones.

The first system, which was adopted in China, India, and probably in pre-Columbian America, offers five modes of playing the scale because, according to the starting note, there are five possible positions for the minor thirds within the octave, characterizing the five different modes:

C D – F G A – C

D – F G A – C D

F G A – C D – F

G A – C D – F G

A – C D – F G A

Similarly, the second system, which was adopted in India, Chaldea, Greece and then across Europe, offers seven modes of playing the scale, with seven possible positions for the semitones E – F and B – C. The Greeks considered each of these modes as capable of arousing a particular emotion (happy, funeral, pompous, etc.).

Incidentally, the description of the heptatonic scale allows us to explain the names of the intervals that we have been using throughout this chapter. European languages have names for only seven notes, the five other ones are described using alterations (F *sharp* or E *flat*, for example). The interval between two consecutive notes of the heptatonic scale is called *second.* It contains one tone (major second: C – D) or one semitone (minor second: E – F). The other names correspond to the notes to be counted in the interval, including the ends. So, C – E is a *third* (C, D, E) and C – G is a *fifth* (C, D, E, F, G). A third is called *major* when it contains two tones (C – E) and *minor* when it contains one tone and one semitone (D – F). With alterations, *augmented* intervals can be formed (as the fourth C – F #) so as *diminished* ones (as the fourth C # – F).

1.2. The tonal system

In a 1962 paper, collected in the book *Music, Architecture*, Xenakis [XEN 71] offered a very useful distinction for understanding the systems of musical composition, and particularly the tonal system. He distinguished between:

– an "out-of-time" structure (relations and operations that are independent of time);

– a "temporal algebra" (sound events forming a set of durations);

– a "within-time" structure (correspondence between the previous two).

The composition system that Greeks left to Byzantines and Arabs was essentially an out-of-time structure, with highly developed and subtle frequency scales, formed by a large number of intervals. Instead, by their aspiration to develop polyphonic music, composers of medieval Europe sacrificed all these subtleties and moved to a within-time structure characterized by harmony and counterpoint.

Under the rule of the modal system, until the early 17th Century, they only used the seven modes of the heptatonic scale constructed with tones and semitones.

To enter the polyphonic reasoning, let us consider the work of Thomas Tallis (*Spem in Allium*, 1573), written for 40 voices divided into eight choirs. The composer Luigi Nono wrote: "with eight choirs, you can only use the tonic and dominant. But the extraordinary thing is that, with this presumed reduction of possibilities, Tallis explores space, makes living space, makes the space itself become *cantabile*" [NON 93].

Starting from an arbitrary note that we can consider as the ground of the harmonic construction, and therefore called *tonic*, we ascend by the interval of a fifth to a second note, which is called *dominant* because it is the second most important note of the scale composed on the tonic: the dominant and the tonic form together a fifth, the most consonant among all chords of two sounds (the octave is considered as an identifier interval).

Let us recall now that the different modes of the scale are not distinguished by the pitch of their tonic, but only by the position of the two semitones in the scale. In the C mode, for example, they are situated between the third and the fourth note (E – F) and between the seventh and the eighth note (B – D). This mode can be played, for example, from the G tonic if we raise the note F by a semitone (F #). We check that halftones are then situated again in the third (B – C) and seventh (F # – G) positions. Therefore, the pitches are relative (we could also change the diapason, which was very fluctuating at the time), and only the intervals are important.

Now, if we form a chord with only two notes, any tonic and its dominant, we cannot tell to which mode it belongs because in all modes there is the same perfect fifth interval (three tones and one semitone), except in the B mode, because the notes B and F form a *diminished* fifth, which encloses the two semitones (B – C D E – F). This is the reason why the modal system rejected the B mode and only worked with six modes.

Except in the extreme case of a work like *Spem in Allium*, it is fitting to enrich the chords in order to enjoy the variety of modes. So, the *mediant* is added, a third above the tonic. There are, therefore, two varieties. In the case of the C mode, the mediant is a major third above the tonic (i.e. two tones) and forms a major chord (C – E – G – C: major third, perfect fifth and octave). The same happens in G and F modes. This is the most consonant three note chord, as it reproduces the harmonic structure of the periodic sound. For the A mode, however, the mediant is a minor third above the tonic (i.e. a tone and a half) and forms the minor chord (A – C – E – A: minor third, perfect fifth and octave). The same applies to the D and E modes.

Polyphony thus engenders a dualism, an opposition between two families of modes, depending on the major or minor chord formed upon the tonic. If we examine the first part of the scale, between tonic and dominant, we see that the C and G modes rise in the same way (a tone, a tone, a semitone and a tone). The same applies to the A and D modes (a tone, a semitone, a tone and a tone). In contrast, the remaining two modes are very peculiar: E mode is the only one that starts with a semitone (E – F), while the F mode begins with three consecutive tones, forming an augmented fourth (F – B). Both modes were pretty well used during the modal time before falling into disuse in the dawn of the tonal revolution.

In the early 17th Century, one can speak of a revolution in music, with its precursory signs (madrigalism, development of new instruments such as the violin, etc.) and then the avalanche of reforms: almost with the simultaneous appearance of opera, concerto with a soloist, continuo, etc., the music became more sensationalist due to a more effective composition system.

The tonal system came from a simplification of the modes, which consisted of strengthening the major/minor dualism already present in the modal system: with only two modes, this dualism can be expressed in a more contrasted, more expressive way.

It can be shown that the modes of C and A are the most different from each other, and they complement each other wonderfully. They were the only survivors, and, thereafter, they were called the *major mode* and the *minor mode*, respectively.

To transpose means to change tonic in the middle of a piece, i.e. the same scale moves up or down. *To modulate* means to change mode, with the same tonic, i.e. switching from major to minor or *vice versa*. These are the operations allowed by the tonal system. We can check that in the midst of a transposition, it is possible to show all the colors of the modes sacrificed in favor of tonal efficiency, and therefore, we have not lost anything consistent with the reduction to two modes, earning in contrast and efficiency. However, at this juncture, what Xenakis called an out-of-time structure changed into a within-time structure: the variety of the seven modes are now found only in the dynamics of transpositions and modulations, subjected to strict rules that allow the harmonic tonal system to work.

In a piece of music with numerous transpositions and modulations, the problem for the composer is to ensure that the listener knows what key it is so that he can follow the "intrigue", without which the whole system loses its interest.

If we play the natural scale (i.e. unaltered) from E or F, even the tonal heard of our time will perceive a somewhat strange melody, something medieval, because, in such a simple case, the tonal system does not have the resources to impose its law. In a simple monody, anything can happen. Now, if we create a melody with the same notes, but where C often returns, insistently, and if we finish with C, undoubtedly we have a melody in C major. But nothing prevents, amid this tune, to do a little foray into A minor, or, with the appropriate alterations, a sudden jump in G major or D minor.

Let us now go to the harmony, the construction of chords. A major chord built on C can adapt to many circumstances: C can certainly be the tonic of C major, but it may also be the mediant of A minor or the subtonic of G major, without any change. To establish the tonality, we need something more: a second chord preceding the first one, which should be so typical, and yet so unstable, as to indicate manifestly, while imperatively needing, its resolution in the first one. Therefore, we will build this second chord on the second most important note of C major, the dominant G. But the chord G – B – D is also a major chord, as stable as C – E – G. Then we will make a chord of four sounds, adding the seventh F and, even more, we will ascend it by a semitone (F #). We get a dominant seventh chord G – B – D – F# perfectly unstable because it contains a semitone (F # – G), the most dissonant interval of the tempered scale. This tension needs a relaxation, and the most ignorant tonal heard feels where it needs to go: to the chord C – E – G – C.

With this formula, we have established the key: hearing this succession of two chords, everyone feels that we are in C major. So we have a reference level: C is on the ground, in level I. Other notes are now tightly ranked; they have their level based on this ground: G is on the fifth floor, level V; E is on the third floor, level III; B is on the seventh floor, level VII. The succession of the two chords on G resolving to C can then be noted V – I: this is the perfect cadence, so called because it always ends a piece in C major, ensuring its downfall. We can then develop this cadence, for example, with a succession III – IV – V – I: so we create a harmonic progression.

Harmony is the discipline of music that describes the constitution of the chords. But it can only do so while explaining the possible progressions, which justify its features. In the tonal system, the morphology cannot be separated from the syntax. The function of the dominant seventh chord is the only explanation of why its seventh has been ascended. In the term "dominant seventh", "seventh" refers to the formation of a four-note chord and "dominant" refers to its role within the tonal progression. Both words have to be juxtaposed to describe the chord. The out-of-time structure (chords) and the temporal algebra (progression) are inevitably linked.

The tonal system is built on two basic principles: the distinction between consonant and dissonant intervals, which dates back to Aristoxenus and the Pythagoreans, and then the strong hierarchy of notes, typical of medieval polyphony, which constantly creates absolute islands within the relative sea of frequency perception.

The tonal system, as its modal ancestor, is then a within-time structure. This is the result of the formidable yearning for polyphony in medieval Europe, unique in the musical history of mankind or humankind.

1.3. Nomenclature of the projections

Today, the central projection is defined from a point source *S*, also known as the center of projection, from which depart the projecting rays, which are half-lines involved in the formation of the image if they intercept both the object to be projected and the board, the other two loci that characterize the projection.

Before explaining what this relatively recent conception offers, and how it differs from the more classic one, which was that of Monge, we must examine the various forms of possible projections and determine the terms that we will reserve to their nomenclature because the expressions used in most books of geometry often cover several meanings and cause a lot of confusion [BEC 03].

Parallel projection can be defined as a central projection where the center of projection is rejected to infinity. We can also define the parallel projection from *a projection plane*, assumed to be infinite, but placed at a finite distance from the object, which emits projecting rays parallel to its normal. This conception better integrates the general description of the projections from the three loci and highlights the analogy with "plane waves" of physics, which have the same configuration.

Similarly, the rays of the central projection are similar to spherical waves and, by analogy with the cylindrical waves, a third type of projection, which we call "axial", is naturally introduced. It will be mentioned in Chapter 3 (see Lambert equivalent cylindrical projection), where the rays are emitted perpendicular to an axis of projection.

The introduction of this *axial projection*, where the emission source is definitely no more punctual, justifies *a posteriori* our reference to a third "locus", which, added to the first two – the one where the object is placed and the projection board – defines the projection itself: it is a *point* for the central projection, a *line* for the axial projection and a *plane* for the parallel projection.

Instead of speaking of *parallel*, *central* or *axial projection*, as we will do here, one would naturally be justified in speaking of "plane", "spherical" or "cylindrical" projection, respectively, if these expressions were not used by most authors in a very different way.

To fully qualify a projection, it is necessary to describe not only the emission source shape but also the board. This is usually a plane, but sometimes one also projects on the sphere, on the cylinder or on the cone. In most of the books, the terms "plane projection" and "spherical projection" refer precisely to the shape of the board.

The term "cylindrical projection" is even more ambiguous. Indeed, it is justified for both the axial emission and the cylindrical board; it is also used to name the parallel projection, as opposed to the central projection, itself described as "conical".

To avoid confusion, we will only the following expressions.

We will only speak of *parallel*, *central* or *axial projection*, *on the plane*, *on the sphere*, *on the cylinder* or *on the cone*, reserving the use of the adjective to describe the type of projection and the circumstantial complement to describe the shape of the board.

Once this nomenclature is fixed, we must justify such expressions as "emission source" or "board interception", which are usual in the ray tracing algorithms. The geometrical methods were developed long before the rise of computer technology, and they found a renewed interest in the study of the sound field and the simulation of visual scenes.

One should not confuse this comeback with the discovery of the dual nature – wave and particle – of light because light particles postulated by these methods of resolution are not photons. On the contrary, as their sound equivalents in acoustics, they are virtual entities that simulate the geometric behavior of involved waves, such as specular reflection, and such behavior appears only when the dimensions are considered to be far superior to wavelengths.

Anyway, the problem of the nature of light does not occur *a priori* in the theories of the projection, since the latter is defined as a transformation from one space to another, and can therefore correctly be regarded as a simple geometrical construction without temporal aspect.

Thus, according to the definition of the projections that we now refer to as "non-oriented", the rays of the central perspective, for example, are considered as infinite lines intersecting at the center of projection (not half-lines departing from the center), and we then study their possible intersections with the object and the board.

In parallel projections, the distinction between non-oriented and oriented definitions is of little consequence, except in the Monge method, which is irrevocably non-oriented. This is consistent with its rigorously geometric and anti-perspective nature.

The non-oriented definition introduces topological breaks, i.e. adjacent points on the object to be projected are no longer adjacent in the projection, and *vice versa*. These features, perfectly correct in terms of projection, of course become strange in the application of perspective drawing. However, the old theories often slipped upon this problem because in hand drawing, a simple test of common sense can remove them, and even eliminate the hidden parts.

However, this does not hold true in the digital resolutions, where it is imperative to use an oriented definition of the projection, which directly removes these problems.

This is why, later on, we always choose this option, after having briefly outlined the features introduced by the non-oriented definition.

The oriented version is that of the projector, the drawings on tracing paper and also the camera: it works with half-lines and can only project half of the three-dimensional space (in front of or behind the vanishing plane, depending on the definition). The non-oriented version, whose projecting lines are infinite, projects the infinite space (except the frontal plane containing the center of projection), and does not match any feasible device: it is the perspective of the geometers of the 19th Century, with its formal elegance and topological ruptures. The oriented version is, inevitably, temporal (the rays are traced); the non-oriented version is necessarily out-of-time.

The selection of the oriented version allows us to unify the formulation of projections with the geometric methods applied to the study of energy fields. In this spirit, we will take all the forms described here, where the projecting rays are straight, under *linear projection* designation (by analogy with, for example, the "linear acoustic"); non-linear forms will be introduced in Chapter 3. This unified formulation makes sense in algorithmic applications, where we see that the paths taken by the sound or light particles are also rays that draw pictures and perspectives.

1.4. The central projection on the plane

Well used, the parallel projection already gives an effect of *depth* on the plane of the sketch, which is the very definition of perspective. The central projection adds a distance effect because the size of objects on the image decreases with their distance from the board.

This strengthening of perspective illusion, added to the planar arrangement of the board, makes the central projection on the plane the simplest and most effective method for the design of pure representation. Since the Renaissance, when artists established its rules, it produced millions of images. Photography, which provides an automatic and immediate realization, has further increased its absolute hegemony in the formation of our visual experience; so we willingly assimilate it to the notion of representation.

This is why, even if the distinction between projection and perspective is particularly instructive here, they quickly pushed toward one another because very abstract constructions in terms of projection often become visually evident in the drawing.

Therefore, in this chapter where the emphasis is on projection and its constructions, we allow a few incursions into the field of view, starting with a few expressions of its vocabulary.

Thus, if the identification of the projection center with the eyes of the observer is incorrect in the non-oriented projection definition, it becomes so natural and illuminating in the oriented definition that it is adopted right now.

A characteristic of the central projection on the plane, an immediate consequence of the shape of the board, is that the distance is featured on the image not with respect to the eyes, but compared to the board. It is only in a perspective sense that we can call this feature an "aberration". In fact, it is fair to say that the perspective induced by this projection matches the visual experience only for relatively small angles just as the parallel perspective coincides with the latter when the dimensions of the shown object are much lower than the distance between the eyes and the object.

Both statements are unequivocal as long as we do not try to quantify too exactly. Therefore, it is dangerous to pretend to measure the drawing "aberrations" found there because they actually depend on the visual experience of the observer, which varies from one individual to another.

Thus, an image may seem absurd to an unwary observer, while another, more accustomed to the method used, understands it at first glance.

1.4.1. *Principle*

The system of central projections consists of projecting a geometric object on the board τ by means of converging rays (Figure 1.2).

The projection center S can be considered as the eye of an observer looking at the object and which receives visual rays from all the points of the object. Therefore, this point of convergence is also called *viewpoint* or *eye*.

The projecting lines are *visual* or *projective rays* constituting a kind of cone or pyramid. The projection plane forming the plane of the drawing is the *picture plane*, *board* or *perspective plane*.

The projection of a point A is the breakthrough point A_p in τ of a ray coming from S. This point A_p is called the *projection*, *image* or *perspective* of A.

The position of S relative to the board τ is often determined by its orthogonal projection P on τ called *principal* or *perspective point* and its distance is called *principal distance*. The circle of center P and radius d drawn on τ is called *distance circle*.

A plane parallel to the board is called *frontal*. We immediately observe that, for any figure included in such a plane, the projection is reduced to a scaling: the image appears on the chart *without distortion*, and it reproduces the figure to a scale factor.

The frontal plane ε passing through S is the *vanishing plane*. The image of a point located in the vanishing plane is at infinity.

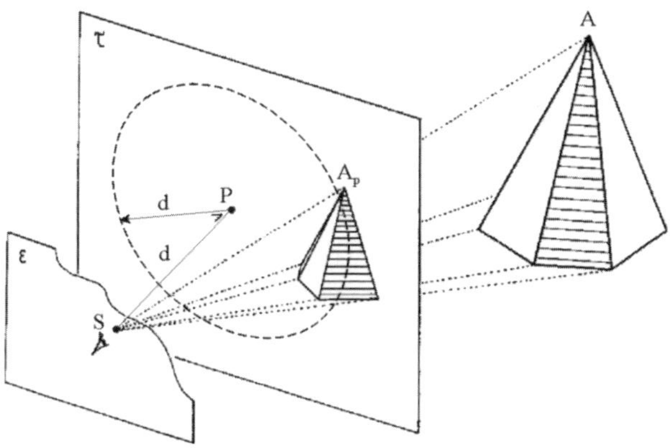

Figure 1.2. *The three loci of a perspective (eye, board and scene)*

1.4.2. *Essential properties*

The central projection is particularly sensitive to the arrangement of the three loci that define it: the eye, the object and the board.

1.4.2.1. *General layout*

These loci may be spread in three different configurations.

The board can be located behind the object τ_1, between the object and the center of projection τ_2, or behind the center of projection τ_3. The first case corresponds to the projected shadow onto a canvas by a light source, while the second is comparable to a decal of a landscape on a translucent paper bonded to a window.

However, the projection is not affected by these arrangements, and it is irrelevant whether the object is placed behind or in front of the board, or even passing through.

In contrast, the third case, which represents the configuration of the eye and the photographic cell, causes an inversion of the image, which must then be recovered: the brain unconsciously performs this operation. However, this configuration disappears in the oriented definition, since the board, if placed behind the vanishing plane, is never reached by the half-lines that intercept the object.

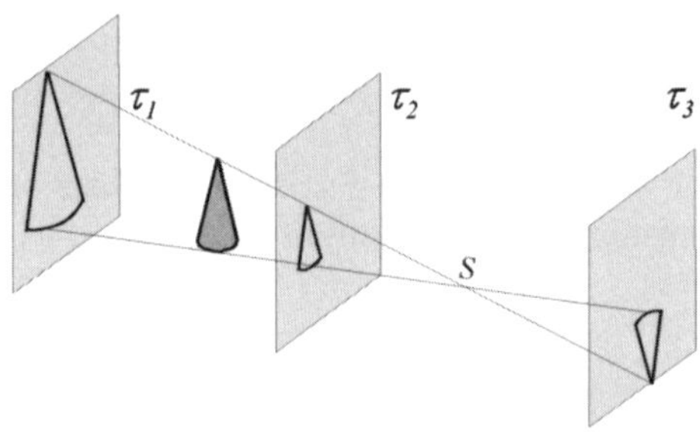

Figure 1.3. *General positions*

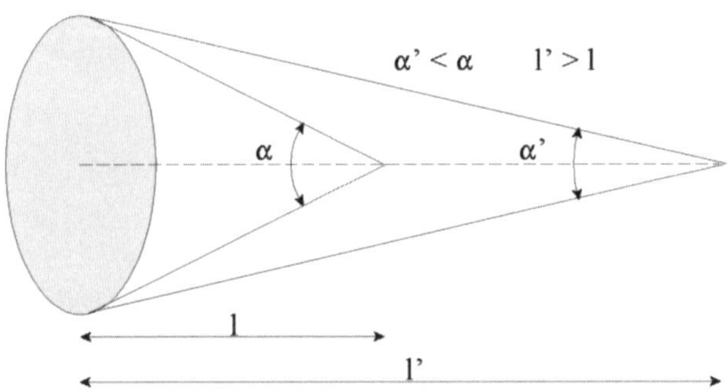

Figure 1.4. *The eye and the object*

1.4.2.2. *The eye and the object*

The opening angle is defined as the largest angle between two rays intercepting the object, i.e. the opening of the smaller cone encompassing the object (Figure 1.4.). If the eye moves away, the viewing angle decreases regardless of the location of the board. Piero de la Francesca recommends using an angle of 60° [FRA 84].

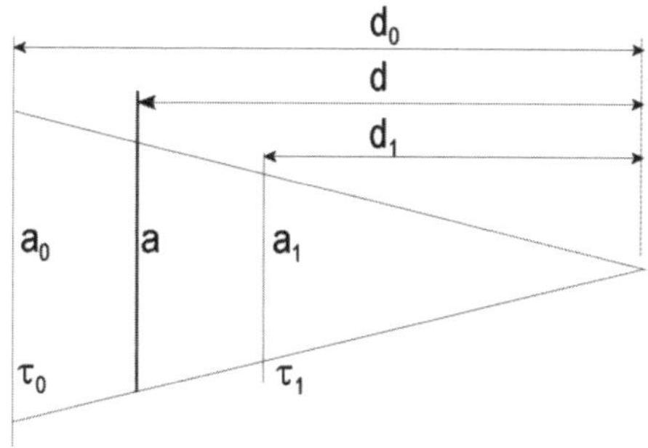

Figure 1.5. *The eye and the board*

1.4.2.3. *The eye and the board*

The eye and the object being fixed, the variation of the distance from the eye to the board, generated by a displacement parallel to itself, induces by Thales' theorem a simple rescaling of the image. The dimensions of the images are related to the distance from the board to the eye by the following relations:

$$\frac{a_1}{a}=\frac{d_1}{d};\quad \frac{a_0}{a}=\frac{d_0}{d};\quad \frac{a_0}{a_1}=\frac{d_0}{d_1} \qquad [1.1]$$

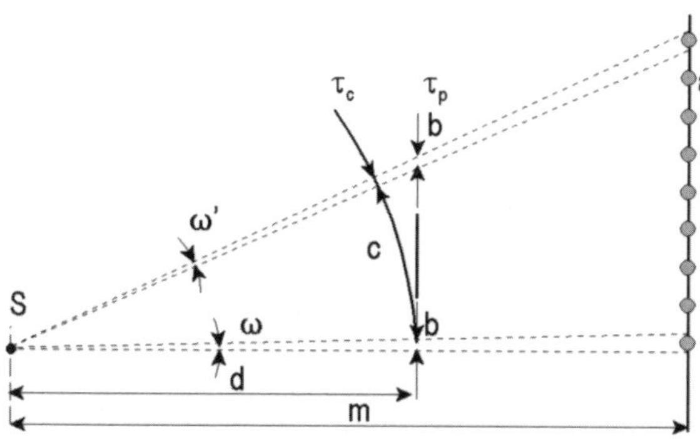

Figure 1.6. *The board and the object*

1.4.2.4. *The board and the object*

A board is placed in parallel to an alignment of columns of diameter *a*. Their images by central projection are assigned the same scale factor: they all have the same width *b* in the drawing regardless of their distance from the center of projection.

This is what painters have called, from this classic example, "the aberration of the central perspective" to mean that the perspective is different in this case of the visual perception, where the distance effect arises simply from the distance to the eye ($\omega' < \omega$, $c < b$).

However, in the central projection, the ratio of dimensions of elements of the object located on the same frontal plane is preserved on the image of this object.

In other words, the effect of distance is represented here not in relation to the eye but compared to the vanishing plane, i.e. to a constant, the board, according to the relation.

$$\frac{b}{a} = \frac{d}{m} \qquad [1.2]$$

1.4.2.5. *Conclusion*

To a scale factor, the image of a central projection on the plane is fully determined by the position of the eye and the observation direction, which is normal to the board.

1.4.3. *Basics*

1.4.3.1. *The straight line*

The perspective of an isolated point is ambiguous because all other points on the straight line that link them to the viewpoint have the same image. One point must always be identified by its position on a line or a plane. It follows that the base of the representation is the straight line perspective.

Let us consider a line in space, the board τ and the viewpoint S (Figure 1.7). The perspective projection or image of the line in the board is defined by two specific points. The first point coincides with its own perspective; this is the trace T_a of the line in the board.

The second point is the perspective of the point at infinity of the straight line, which is the meeting point with the board of the perspective radius led by S parallel to the given line. This point is denoted by F_a in the case of the line a. F_a is the vanishing point of the line a; SF_a is the vanishing radius. The perspective of a straight line is therefore T_aF_a, but this perspective also extends upstream of the trace and downstream of the vanishing point.

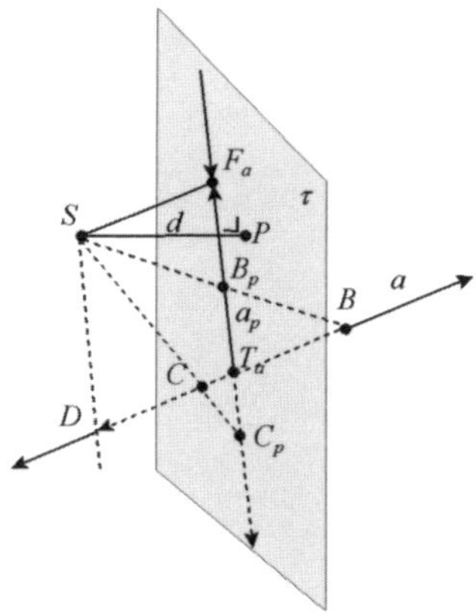

Figure 1.7. *Image of the straight line*

The T_aF_a part corresponds to all points such as B behind the board. All of the upstream to T_a portion corresponds to points such as C located in front of the board between T_a and the vanishing plane. The last point, D, is the intersection of the line a and the perspective ray SD parallel to its perspective a_p. It belongs to the vanishing plane and its image is located at infinity. Part of the perspective downstream of F_a, which corresponds to points located in front of the board, from D to infinity, and introduces a topological break, disappears in the oriented definition. In this case, the image of an infinite straight line becomes a half-line interrupted at the vanishing point.

In the plane of the drawing, we get images of lines a, b and c shown with their trace and their vanishing point (Figure 1.8).

Note that the vanishing points depend mainly on the position of the viewpoint. This position is defined by the position in the plane of the perspective drawing of point P and the knowledge of the principal distance d. The distance circle on the working drawing provides both. The classic drawing of a perspective sketch always starts by the distance circle or at least its center P, d is assumed to be known. The vanishing point of a line characterizes its direction while its position is defined by its trace in the board.

1.4.3.2. *The plane*

A plane is represented by its trace t and its vanishing line f. The trace is the intersection of the plane with the board (Figure 1.9); *this is the locus of the traces of all lines in the plane*. The vanishing line is the intersection with the board of its vanishing plane, which is drawn through the viewpoint parallel to the given plane; *the vanishing line of a plane is the locus of vanishing points of any line that it contains. The lines t and f of a plane are, by definition, parallel*, plane α is any plane at all.

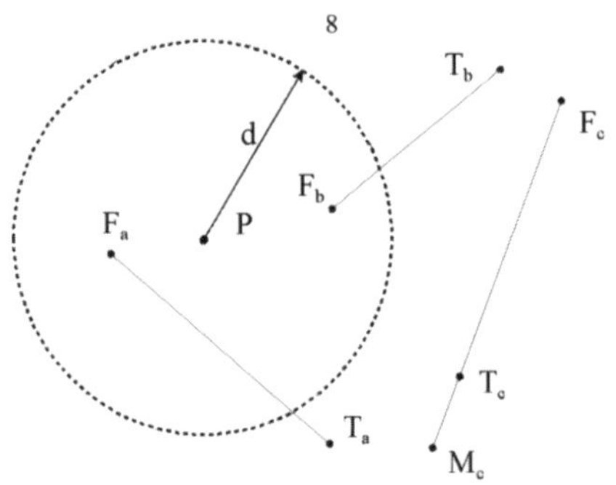

Figure 1.8. *Image of the line on the drawing*

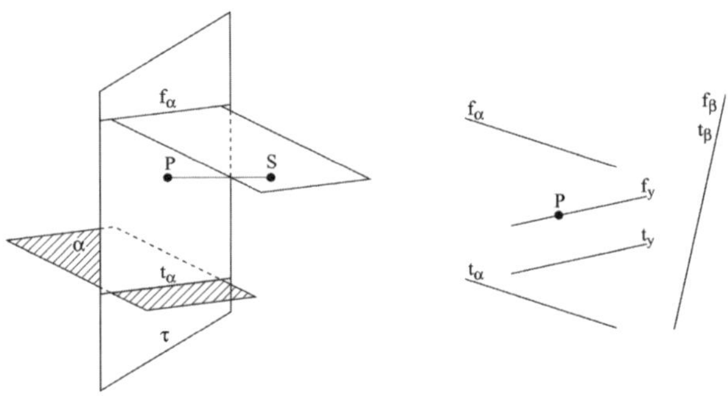

Figure 1.9. *Image of the plane*

As the vanishing point represents the direction of the lines, the direction of a plane is characterized by its vanishing line, which is unique for all parallel planes (Figure 1.10).

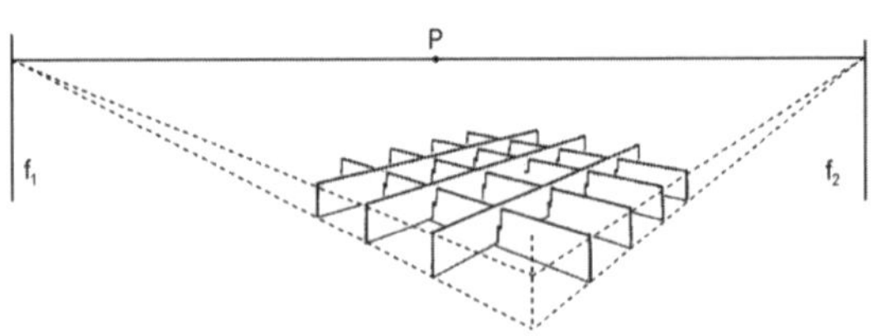

Figure 1.10. *Vanishing lines*

1.4.3.3. *The point*

The perspective of a point can be defined by a single point in the board; the point is indeed the image of all the points of a projecting ray (Figure 1.11). The perspective of a point must be fixed on the perspective of a line or a plane. Thus, the point M_1 is defined if and only if it is positioned on the line T_1F_1; so we have to include this information in a comment in the drawing, or to introduce a special conventional notation (e.g. by means of a common index to all points of the line).

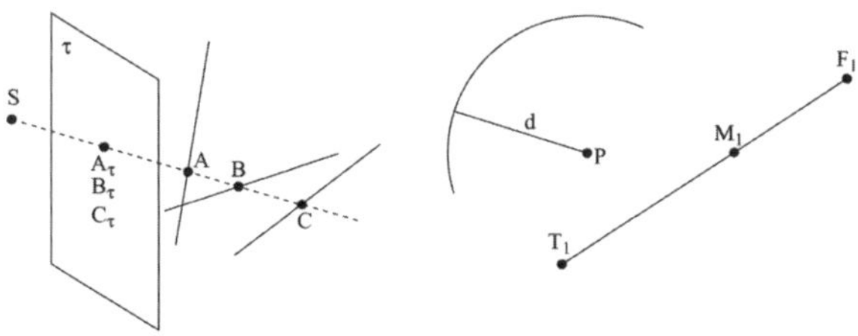

Figure 1.11. *Images of points*

1.5. Proportions and progressions

In the document “The collection” of Pappus of Alexandria [DAL 33, BOR 86], we find the definition of the arithmetic, geometric and harmonic averages:

$$\frac{a+b}{2}; \ \sqrt{ab}; \ \frac{2ab}{a+b} \qquad [1.3]$$

These averages are directly related to the series or progressions of the same name.

Figure 1.12. *Arithmetic progression*

1.5.1. *Arithmetic progression: AB = CD =...*

Equidistant points ($AB = BC = CD$...) naturally form arithmetic progressions. This growth has neither beginning nor end: it extends to infinity equally in both directions.

The distance between two points on the arithmetic progression is usually referred to as “ratio”, but here this term does not have the meaning given by Aristotle and

Euclid. This is why we prefer to call it *step*: the step of an arithmetic progression is the difference p between the positions of two successive points.

We may add that the progression has two *poles* located at $-\infty$ and $+\infty$, which are unattainable and the same for all progressions. One will talk to the arithmetic progression of two *poles at infinity*.

We can also define the arithmetic progression with a point (for example A) and the step p, which is any number, positive or negative but different from zero.

$$A \pm np \quad , \quad -\infty \leq n \leq \infty \qquad [1.4]$$

1.5.2. *Geometric progression:* $\frac{BC}{AB} = \frac{CD}{BC} = r$

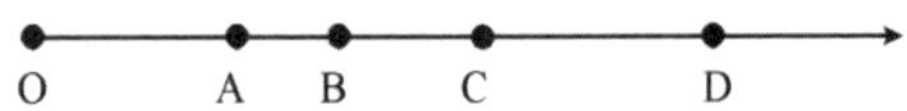

Figure 1.13. *Geometric progression*

Let us consider four collinear points A, B, C and D such that their intervals are increasing: AB < BC < CD:

$$\frac{BC}{AB} = \frac{CD}{BC} = r; \; r > 1 \qquad [1.5]$$

Points A, B, C and D are marked with respect to an origin point which is such that: $\frac{OB}{OA} = \frac{B}{A} = r$

This definition can also be written as:

$$\frac{B}{A} = \frac{A + AB}{A} = r \Rightarrow A = \frac{AB}{r-1} \qquad [1.6]$$

Then we check that:

$$\frac{C}{B} = \frac{A + AB + BC}{A + AB} = 1 + \frac{BC}{AB\left(1 + \frac{1}{r-1}\right)} = 1 + r - 1 = r \qquad [1.7]$$

The geometric progression can be written in two ways:

$$\frac{BC}{AB} = \frac{CD}{BC} = \frac{DE}{CD} = \cdots = r \qquad [1.8]$$

$$\frac{B}{A} = \frac{C}{B} = \frac{D}{C} = \frac{E}{D} = \cdots = r \qquad [1.9]$$

Browsing the progression to the right, each interval is equal to the previous one multiplied by r. Since $r > 1$, the progression tends to $+\infty$. Browsing the progression to the left, each interval is equal to the previous one divided by r. Since $r > 1$, the intervals tend to zero and the points move toward the origin, the second limit of the progression.

Therefore, the increase takes place between the two poles, the origin (O) and the infinite ($+\infty$). If the ratio is less than 1, we obtain the inverse progression, identical, but in the other direction. Each geometric progression so determines its own pole O, a reference point for its expansion, zero of its arithmetic expression and the limit of its spatial extension.

In the following examples, the geometrical progressions are defined by two successive intervals; they include two poles, in 0 and at infinity. From [1.5] and [1.6], the ratio and the position of the point A are computed.

$AB = 1$	$BC = 2$	$\Rightarrow$	$r = 2$	, $A = 1$	series: ..., 1, 2, 4, 8, 16, ...
$AB = 2$	$BC = 5$	$\Rightarrow$	$r = 5/2$	, $A = 4/3$	series: ..., 4/3, 10/3, 25/3, ...
$AB = 1$	$BC = 1.5$	$\Rightarrow$	$r = 1.5$	, $A = 2$	series :..., 2, 3, 4.5, 6.75, 10.125,
$AB = 1.5$	$BC = 2.5$	$\Rightarrow$	$r = 5/3$	, $A = 9/4$	series: ..., 9/4,15/4, 25/4,125/12,

We can also describe the progression by a point and the interval which follows or precedes it. In this case, equation [1.6] allows us to calculate the ratio, and any other point directly.

$AB = 1$	$A = 1$	$\Rightarrow$	$r = 2$	, series: ..., 0.5, 1, 2, 4, 8, 16, ...
$AB = 2$	$A = 1$	$\Rightarrow$	$r = 3$	, series: ..., 1/3, 1, 3, 9, 27, ...
$AB = .5$	$A = 1$	$\Rightarrow$	$r = 1.5$	, series: ..., 2/3, 1, 1.5, 2.25, 3.375, ...
$AB = -.8$	$A = 1$	$\Rightarrow$	$r = .2$	, series: ..., 5, 1, .2, .04, .008, ...

1.5.3. *Harmonic progression:* $AB = BC = CD, \ \frac{1}{A}, \frac{1}{B}, \frac{1}{C}$

Let us consider four collinear points A, B, C and D such that their intervals are increasing: $AB < BC < CD$.

Figure 1.14. *Harmonic progression*

A harmonic progression is composed of terms that are the inverse of an arithmetical progression. Take, for example, a decreasing arithmetic series of step p.

$$x+p \quad x \quad x-p \tag{1.10}$$

The corresponding harmonic series is increasing.

$$\frac{1}{x+p} \quad \frac{1}{x} \quad \frac{1}{x-p} \tag{1.11}$$

It is based on the positions *A*, *B* and *C* measured from any origin *O*.

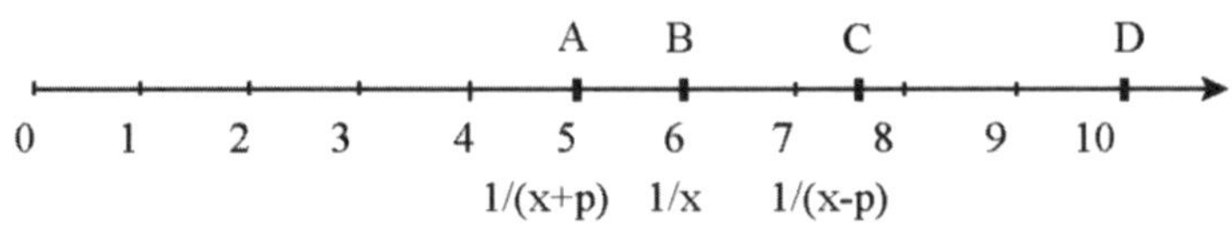

Figure 1.15. *Details of the harmonic progression*

The values of *x* and *p* are deduced according to the intervals of the images of *AB* and *BC*.

$$\begin{aligned} AB &= \frac{1}{x} - \frac{1}{(x+p)} \\ BC &= \frac{1}{(x-p)} - \frac{1}{x} \end{aligned} \tag{1.12}$$

Then:

$$\begin{aligned} x &= \frac{BC-AB}{2 \cdot AB \cdot BC} \\ p &= \frac{(BC-AB)^2}{2 \cdot AB \cdot BC \cdot (AB+BC)} \end{aligned} \tag{1.13}$$

By choosing $AB = 1$ and $BC = 1.5$ (Figure 1.15), we obtain:

$$x = \frac{1}{6} \quad p = \frac{1}{30} \qquad [1.14]$$

When the arithmetic series tends to infinity, the corresponding harmonic series progresses toward a pole of coordinate $1/\infty$, i.e. zero. The other end of the harmonic progression is obtained when $x - nr = 0$, i.e. if the corresponding arithmetic series passes through the origin. In the example above, this value is obtained for the point G of the series A, B, C, D, E, F and G. In equation [1.15], the first line contains the terms of an arithmetic progression and the last line contains the corresponding terms of the harmonic progression.

$$\begin{array}{ccccccc} \frac{6}{30} & \frac{5}{30} & \frac{4}{30} & \frac{3}{30} & \frac{2}{30} & \frac{1}{30} & 0 \\ A & B & C & D & E & F & G \\ 5 & 6 & 7.5 & 10 & 15 & 30 & \infty \end{array} \qquad [1.15]$$

If the arithmetic series extends in negative values, a discontinuity in the corresponding harmonic series occurs, which changes abruptly from a positive value that can be substantial (up to infinity if a node is at the origin) to another value, this time negative (and which can also be quite large if a node is in the negative part, but close to the origin).

This type of behavior reminds us of the central projection of a line in the non-oriented version, where the same topological break from the vanishing point appears, in the same way as it appears here to the *pole at the origin* of the harmonic progression.

1.6. The eighth proposal of Euclid

Euclid's *Optics* is a work on the geometry of vision written around 300 BC. Technically, it is a book about ray trace. For the first time, sight is modeled mathematically by a discretization in visual rays scanning the viewed scene. Due to limitations of the time – the Greeks had not yet acquired trigonometry – all demonstrations are performed by the geometry of the figures, particularly by considering similar triangles. Despite this, this book is a masterpiece that was read a lot during the Renaissance [BEC 05].

In proposition 8, Euclid shows that the perceived size of an object is not related to its distance from the eye by a simple proportion: "The equal and parallel magnitudes located at different distances from the eye are not in proportion to the distances" [EUC 00].

This is serious, and it must have caused a great disappointment to Euclid. According to the geometry, sight would not perceive spatial proportions, unlike hearing with musical ones, and unlike the general intuition expressed by Aristotle in *On the Soul* [ARI 94]:

> If voice always implies a concord, and if the voice and the hearing of it are in one sense one and the same, and if concord always implies a ratio, hearing as well as what is heard must be a ratio. That is why the excess of either the sharp or the flat destroys the hearing. (So also in the case of savors excess destroys the sense of taste, and in the case of colors excessive brightness or darkness destroys the sight, and in the case of smell excess of strength whether in the direction of sweetness or bitterness is destructive.) This shows that the sense is a ratio. That is also why the objects of sense are pleasant when the sensible extremes such as acid or sweet or salt being pure and unmixed are brought into the proper ratio; then they are pleasant: and in general what is blended is more pleasant than the sharp or the flat alone; or, to touch, that which is capable of being either warmed or chilled: the sense and the ratio are identical: while in excess the sensible extremes are painful or destructive.

Aristotle expressed the relativity of perception in general. To the Greeks, this perceptual relativity is the field where the proportions are necessary and sufficient tools to compare magnitudes that do not refer to any absolute value. So, the distances cannot be set in proportions to our eyes.

This discovery is doubly surprising: on the one hand, we are influenced by the Greek theory of proportion and, on the other hand, we know, from the Renaissance, that the proportions that occur in real space objects also appear directly to our eyes.

Euclid attempted to recover some of the visual geometrical proportion (propositions 48 and 49, based on consideration of the capable arcs) that do not subtract anything to the categorical character of his eighth proposition, and there is no error in the elegant demonstration he offers, which is detailed below:

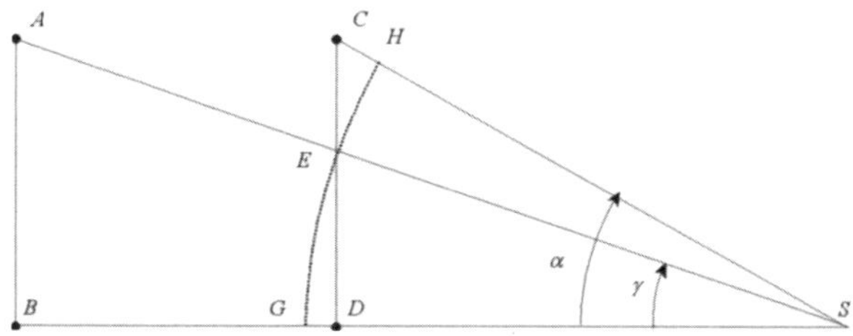

Figure 1.16. *Scheme for Euclid's eighth proposal*

Euclid demonstrated (Figure 1.16) that the ratio of angles γ and α from which the segments AB and CD are seen equal and parallel is not equal to the ratio inverse of their distances from the observation point S:

$$SB / SD > \alpha / \gamma$$

First, the *GEH* arc centered in *S* is drawn.

The proof is purely geometrical, and it only uses surface ratios (we call "area" the area of a triangle and "sector" the area of a circle sector; both entities are defined with three points):

Area *SEC* > Sector *SEH*

Area *SED* < Sector *SEG*

So: Area *SEC* / Sector *SEH* > Area *SED* / Sector *SEG*

And: Area *SEC* / Area *SED* > Sector *SEH* / Sector *SEG*

Summing numerators and denominators in the relations:

(Area *SEC* + Area *SED*) / Area *SED* > (Sector *SEH* + Sector *SEG*) / Sector *SEG*

Now: Area *SCD* / Area *SED* > Sector *SGH* / Sector *SGE*.

For right triangles:

Area *SCD* / Area *SED* = *CD* / *ED*

And by Thales' theorem:

AB / ED = SB / SD

As AB = CD, by hypothesis:

SB / SD = Area SCD / Area SED

So: SB / SD > Sector SGH / Sector SGE

As the sectors are each other as their angles:

$SB / SD > \alpha / \gamma$

Therefore, the ratio of the distances is greater than the ratio of the angles. (QED)

Comparing this demonstration, or any other of Euclid's *Optics*, with the writings of Renaissance perspectivists, an essential difference immediately jumps out: Euclid never talks about a board, he works without any screen, and pictures are formed directly in the eye without intermediary. He just considers the viewing angles. He never projects onto the plane.

Projecting without a board is the same, geometrically, as projecting onto any spherical board centered on the view point (e.g. the one manifested in the above proof by its trace *GEH* in the figure). Indeed, in this case, the projected distances are equal (obviating a scale factor) to the viewing angles, meaning that the magnitudes considered by Euclid are directly measured in the drawing.

This does not mean that Euclid thought in a spherical board, which would be an anachronism, since Euclid did not think at all in a board. However, his reasoning without a board is mathematically equivalent to a projection onto the sphere. This observation allows, including Euclidean optics in the general projection theory, adding a third locus, a neutral one – the spherical board – between the two ones explicitly mentioned by Euclid: the eye locus and the object one. In this sense, there is nothing paradoxical to assert simultaneously that Euclid projected without any board and that his optics model corresponds to a projection onto the sphere.

A possible explanation lies in the *quadrivium* – arithmetic, music, geometry and astronomy – which organized the teaching of science in Europe since antiquity and throughout the Middle Ages. Arithmetic came from music, as we have shown, and geometry from sight. Astronomy is a less easy application. There is no board in the sky, but the *vault of heaven*, a spherical board.

When painters of the Quattrocento applied Euclid's ideas to their plane boards, they immediately found the solution. Indeed, the central projection *onto the plane* of an arithmetic progression is always a harmonic progression, as we will show in the next chapter. Perspective drawing is therefore provided with a mathematical law as precise and legitimate as the one that allowed the building of the musical scale. At first, mathematicians were the only ones left on the margins of the new geometry, and of its development by artists (painters, architects, designers, etc.). They could not give up the synthetic form advocated by Plato and implemented by Euclid in his other great book, the *Elements*. This book can be studied at the beach, drawing figures in the sand, identifying their parts with few letters and then closing the eyes and meditating. And painters now claim to reproduce drawings, abandon abstraction and eternity conquered with difficulty on the natural laziness of men, and start thinking with imperfect eyes, not with the divine soul, sacrificing universality to focus on a miserable world of pigments, fragile and inconstant, desperately flat and openly misleading.

Filippo Brunelleschi (1377–1446) discovered the basic rules of central perspective; Leon Battista Alberti (c. 1404–1472) presented them for the first time in his treatise "Della pintura" in 1435. Piero della Francesca (c. 1416–1492) refined them in "De prospectiva pigendi" (c.1480) and Leonardo da Vinci (1452–1519) recorded his observations in his famous notebooks in the early 1490s.

However, it was not until the reasoning of the French engineer and geometrician Gérard Desargues (1598–1662) that the first purely mathematical study of the central projection became available. He was a friend of René Descartes (1596–1650) and his work, largely lost, has been admired by Blaise Pascal (1623–1662), who studied the conical projection. Later, Gaspard Monge (1746–1818) developed descriptive geometry, that is to say, essentially, the parallel projection onto the dihedral, and Jean-Victor Poncelet (1788–1867) completed the description of projective geometry. Unlike what its name suggests, this is an abstract and difficult discipline, which provides sophisticated methods (anticipated by the work of Pascal) for the solution of complex mechanical problems. These methods have become obsolete with the development of the computer.

Many art historians have argued that the central perspective was the great discovery of occidental culture, a characteristic of European rational effort, by which our civilization is distinguished. However, European art was singled out long before, with the development of music as a within-time structure.

The Renaissance artists would never have felt the importance of perspective rules – to the example of their Greco-Latin models – had it not been for their musical training, which forced them to look at certain correspondences between visual and auditory proportions, hoping to get in visual arts what had already been achieved in music. The great Italian theorists, such as Leonardo da Vinci, admired that the perspective foreshortening was governed by the same ratios that determine musical intervals. In such observations, Renaissance artists assembled their utopia of a world full of sensory correspondences. They just missed the free handling of animated images, where the central perspective is revealed in all its strength and all its meaning.

Despite the subsequent geometrical generalization of perspective by mathematicians, we still feel in all manual drawing, its first development – musical – by painters. The mixture of both purposes, mathematics and art, has generated the greatest confusion. What is the meaning, for example, of the traditional distinction between perspectives with one, two or three vanishing points? For a geometer, none: there are as many vanishing points in the image as directions in the scene. But if we consider the cube as the basic structural element, we can show it from the front (one vanishing point), sideways (two points) or sideways with high-angle shot (three points): we feel in this inspection a remnant of the lost dream of structuring perspective as a musical scale, both out-of-time (basic schemes of composition) and temporarily (eye movement).

The perspectivists addressed visual forms in the same way that musicians were employing sounds: as structural elements to be previously arranged according to the same rules that govern the composition course. Natural forms are mere noise, amorphous, not directly usable. Musical sounds are distinguished by their harmonic structure, which provides them an exact position on the musical scale. The composition rules of the modal system are then built in this same harmonic structure. Well, for the Renaissance painters, and even architects, architecture will have no other function than previously organizing visual material, defining exact positions that serve to form beautiful perspectives, according to the same rules.

Unlike Chinese painting, European art will proceed, thereafter, from architecture to nature, which will also be instrumentalized by landscapers, to "create perspective", i.e. to join the new perspective vision of the world. In the Baroque era, the myth of the perfect match between all the arts, powered by exquisite disquisitions on the golden section and the pure forms of geometry (sphere, ellipse, etc.), will move to the theater where control is absolute. There, a public ideally arranged in a circle (according to the acoustic ideas of the time) comes into contact with the scene, a box of musical and perspectival illusions, dispensing proportions and disproportions, etc.

Finally, Gotthold Ephraim Lessing (1729–1781), in his "Laocoon", inaugural work of contemporary aesthetics, claimed the overwhelming difference between poetry and painting, between the arts of time and the arts of space [LES 90]. Today, if we treat central perspective with neutrality, we must clear its theory of the 19th Century excessive geometric abstraction and, at the same time, of the musical instrumentation bequeathed by the Renaissance, inoperative because too naive: we must, therefore, study its intrinsic – purely geometric – properties.

From the countless perspective representations available today, we know that the properties of the central projection are also the properties of human vision. Studying the first, we will better know the second.

2

Main Properties of Central Projections

2.1. Straight lines and conics

When we look at a photograph or a film, which are direct outputs of the central projection on the plane, we have the illusion of reality, i.e. we find in these perspectives the geometric properties of the visual space. We will now explore the properties of the central perspective, which also structure our visual perception of space.

It may seem counterintuitive, since this perspective distorts many things, starting with the parallels, the proportions and the circles. But it distorts them in a particular way, and our eyes do the same. In our viewing, straight railways seem to converge to infinity, and circular objects appear to us more often as ellipses. However, we feel that these rails are parallel and these ellipses are circles.

Straight lines are an invariant of the central projection on the plane. In Chapter 3, we will see other forms of projection, and other types of boards, where straight lines bend. This gives strange perspectives that seem distorted.

Conics are another invariant of the central perspective, but, in this case, any projected conic curve may be transformed into another one. To our eyes, which always seek to identify the simplest structure, conics are figures of the circle probably distorted by perspective. A vertical circle illuminated by a point source projects an elliptical shadow on the ground (Figure 2.1), but if the circle is tangent to the vanishing plane, then the projected shadow becomes a parabola (Figure 2.2).

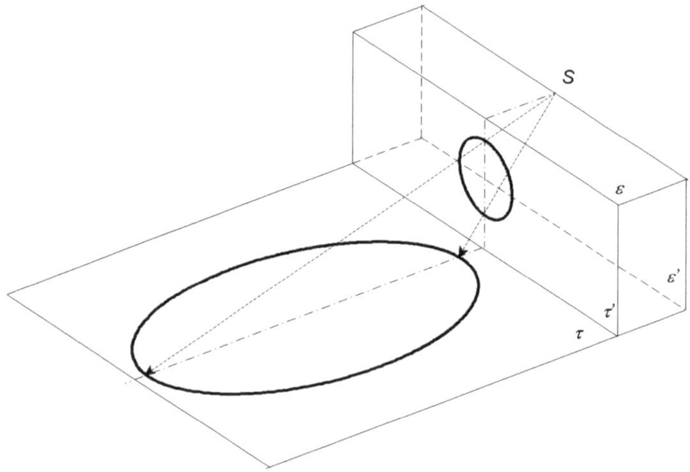

Figure 2.1. *Ellipse and its generator circle*

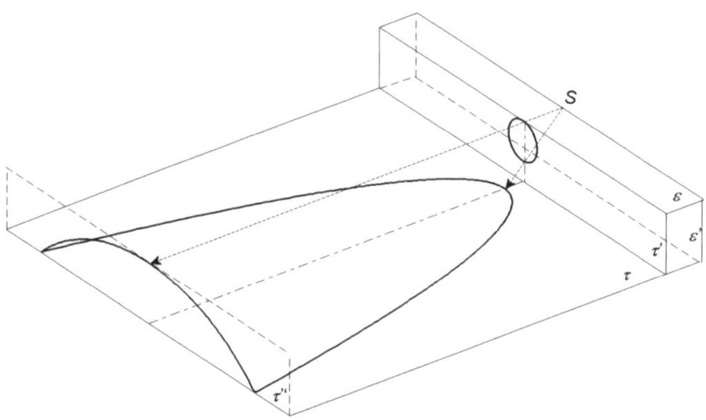

Figure 2.2. *Parabola and its generator circle*

Finally, Figure 2.3 shows the case in which the circle cuts the vanishing plane ε; in the oriented version of the projection, only the part of the circle located under the plane ε can be projected, in the plane forming a branch of a hyperbola τ (left). In the non-oriented version, the image is dissociated: the infinite projecting lines intercepting the top of the circle also intercept the box in the other direction, forming the other branch of the hyperbola (right). In the drawing, the rays close to the vanishing plane have been stopped in the frontal planes τ^1 and τ^2, where they form circular images of the object.

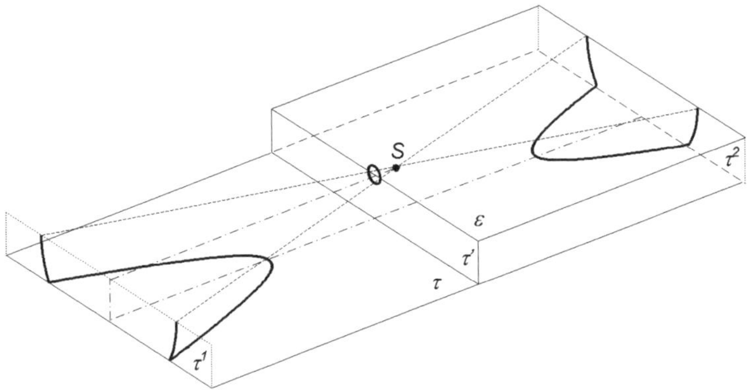

Figure 2.3. *Hyperbola and its generator circle*

This case illustrates the peculiarities of the non-oriented version: the second branch of the hyperbola is a purely virtual one, and it cannot be seen by an eye *S* nor can it be illuminated by a spotlight *S*, but it completes the conic curve. So we can only properly say that a conic curve is always projected as another *whole* conic curve.

2.2. Coherence and cross ratio

In geometry, the cross ratio (or anharmonic relation, or double ratio) is a number associated with an ordered quadruple of four collinear points [BEC 03]:

$$(ABCD) = \frac{CA}{CB} \Big/ \frac{DA}{DB} \qquad [2.1]$$

By interchanging the order of points, 24 combinations are obtained, which can be reduced to six independent ones. These results are also presented in some standard configurations (Table 2.1).

$$\begin{array}{ccccccccc} (ABCD) & = & (BADC) & = & (CDAB) & = & (DCBA) & = & r \\ (ABDC) & = & (BACD) & = & (CDBA) & = & (DCAB) & = & 1/r \\ (ACBD) & = & (BDAC) & = & (CADB) & = & (DBCA) & = & 1-r \\ (ACDB) & = & (BDCA) & = & (CABD) & = & (DBAC) & = & 1/(1-r) \\ (ADBC) & = & (BCAD) & = & (CBDA) & = & (DACB) & = & (r-1)/r \\ (ADCB) & = & (BCDA) & = & (CBAD) & = & (DABC) & = & r/(r-1) \end{array} \qquad [2.2]$$

Harmonic division $r = -1$	Equal segments $r = 4/3$	Equal 60° arcs on a circle $r = 3/2$	Geometric progression (2, 4, 8, 16) $r = 9/7$
−1.0000	1.3333	1.5000	1.2857
−1.0000	0.7500	0.6667	0.7778
2.0000	−0.3333	−0.5000	−0.2857
0.5000	−3.0000	−2.0000	−3.5000
2.0000	0.2500	0.3333	0.2222
0.5000	4.0000	3.0000	4.5000

Table 2.1. *The six cross ratios for different point configurations*

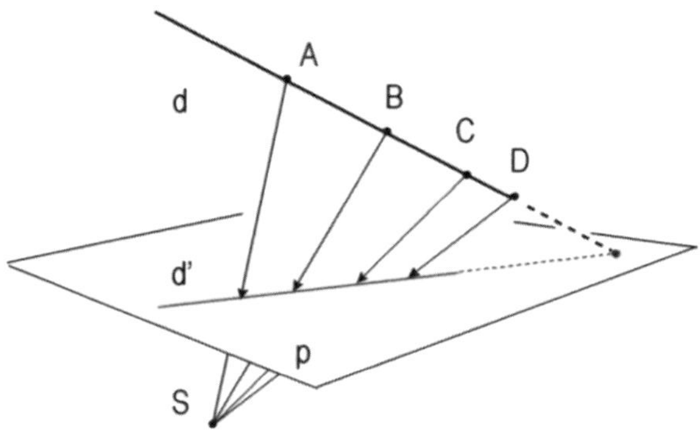

Figure 2.4. *Cross ratio of four collinear points*

The cross ratio of a pencil of lines (Figure 2.5) is calculated in the same way. Denoting by *a*, *b*, *c*, *d* the lines coming from *P* and passing, respectively, by P_1, P_2, P_3, P_4, we obtain the cross ratio of the pencil as a function of the angles in point *P*.

$$(abcd) = \frac{\sin(\alpha_1 + \alpha_2)\sin(\alpha_2 + \alpha_3)}{\sin\alpha_2 \sin(\alpha_1 + \alpha_2 + \alpha_3)} \qquad [2.3]$$

The cross ratio of four lines is equal to the cross ratio of four points obtained by the intersection of the pencil with any line.

Given a straight line isolated from any context, we measure regular intervals, choosing as markers a point of origin *A* and a point *B*. The *AB* segment serves as a reference length. The position of a point *X* on the line is defined by the variable *s* which varies from $-\infty$ to $+\infty$.

$$X = A(1-s) + Bs \qquad [2.4]$$

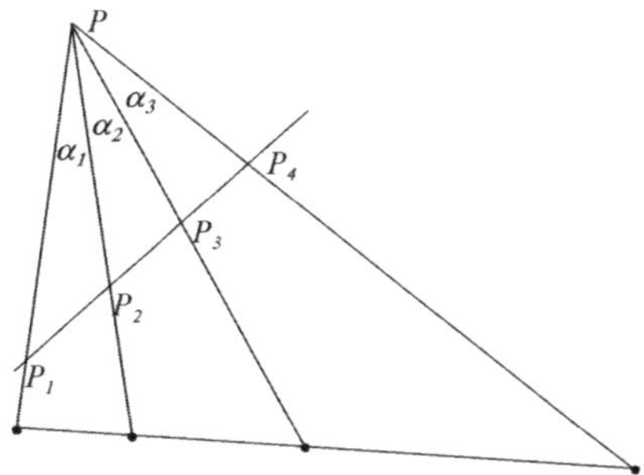

Figure 2.5. *Cross ratio of a pencil*

In this way, we can measure the distance from the origin A with the parameter s. The signed distance from a point X to A is simply s. From the three points A, B, X and the point at infinity P_∞ (defined by $s_P = \infty$), the cross ratio [2.1] is calculated:

$$(A,X,B,P_\infty) = \frac{1(s_P - 1)}{(1-s)s_P} \qquad \text{or} \qquad (A,B,X,P_\infty) = \frac{ss_P}{((s_P-1)(s-1))} \quad [2.5]$$

If s_P is infinity, $s_P/(s_P - 1) = 1$. The cross ratio of these four points is equal to the ratio (ABX) and is -1 if the point X is in the middle of the segment AB.

$$(ABX) = \frac{s-1}{s} \quad [2.6]$$

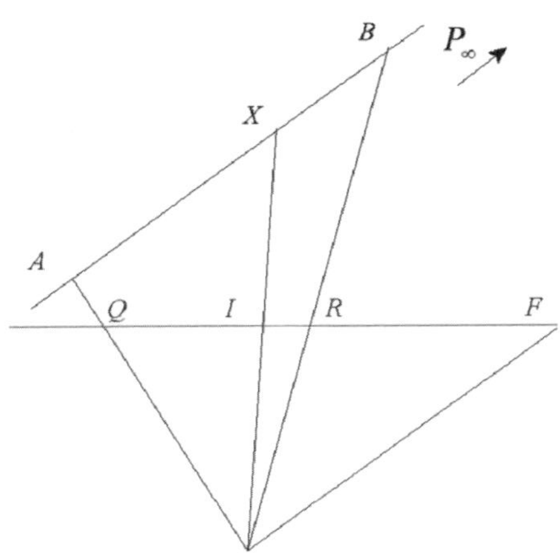

Figure 2.6. *Use of a parametric definition of a point on a line*

In Figure 2.6, the image of the straight line is defined by the images Q of A, R of B and F of the point at infinity P_∞. As the A-B segment does not cross a line parallel to Q-R (equivalent to the vanishing plane), F is outside Q-R. The cross ratio (A B P_∞ X) is known and is equal to the ratio of (A, B, X). We can deduce the position of the image I of X by calculating the parameter s_i that marks it with respect to the segment Q-R.

$$\frac{s_P(s-1)}{(s_P-1)s}=\frac{t_i(s_i-1)}{(t_i-1)s_i} \qquad [2.7]$$

As $s_P = \infty$ and $s_P / (s_P - 1) = 1$, the expression becomes:

$$s_i=\frac{st_i}{s+t_i-1} \qquad [2.8]$$

where t_i indicates the position of the vanishing point F of the line relative to the images Q and R of A and B (t_i must be different from 0 and 1 because neither A nor B is located at infinity) and s_i indicates the position of I with respect to Q and R. The relationship between them is linear rational.

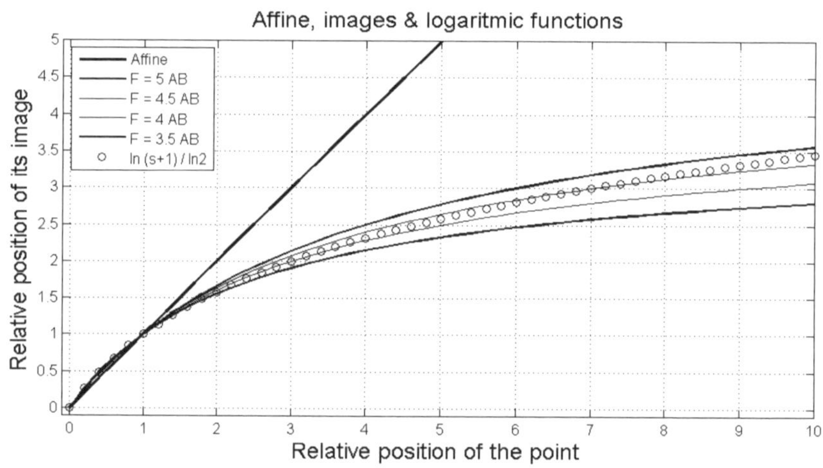

Figure 2.7. *Position of a point on a segment versus its position on the image*

When s tends to infinity, $s_i = t_i$ and I, the image of X, approaches the vanishing point F. If t_i is large, the vanishing point is far from the Q-R segment and distances are measured almost linearly on the image of the right (affine transformation).

$$\lim_{t_i \to \infty} s_i = \lim_{t_i \to \infty} \left[\frac{st_i}{s + t_i - 1} \right] = s \qquad [2.9]$$

The cross ratio of the pencil is calculated according to [2.3] (Figure 2.5 [LAV 13]). If the three angles defining the arcs are equal: $\beta_1 = \beta_2 = \beta_3 = \beta$, the cross ratio is given by:

$$cr = \frac{\sin^2 \beta}{\sin \frac{\beta}{2} \sin \frac{3\beta}{2}} \qquad [2.10]$$

For β = 60°, it is equal to 1.5 regardless of the position of the fifth point from which the lines are drawn. This formula also shows that the cross ratio tends to 4/3 if β is small, as for a line segment divided into three equal parts. Vector products allow us to easily calculate the sinus of formula [2.10]. Thus, for the angle α_1 of Figure 2.5, we have:

$$\sin(\alpha_1) = \frac{(PP_1) \wedge (PP_2)}{|PP_1||PP_2|} \qquad [2.11]$$

2.2.1. *Calculation of cross ratio on a circle*

The cross ratio of four lines constructed on a circle is invariant regardless of the position of the origin point on the circle (we assume that the five points – four at the ends of the lines and one at their intersection – are always situated on the circle). This is easily demonstrated as the capable segments of the three concerned arcs remain unchanged when the top is moved (Figure 2.8).

For the parabola (Figure 2.9), we find the same result as for the circle. The curve in Figure 2.10 is valid for all the conics. One computes the cross ratio for an angular aperture of three equal segments ranging from 2 to 240°. For a 180° angular aperture, it is equal to 1.5. When the angular aperture tends to zero, the cross ratio tends to 4/3, its value for four collinear equidistant points.

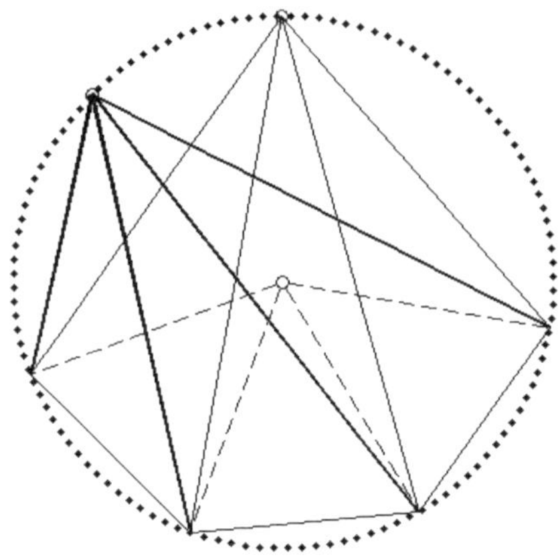

Figure 2.8. *Three 50° arcs. The top angles of the two pencils are identical*

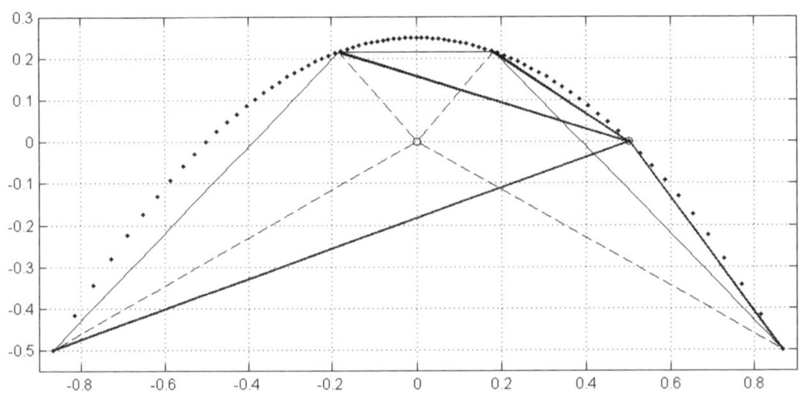

Figure 2.9. *Five points on a parabola for a total sector of 240°*

In the frontal plane, which does not deform anything, circles are circles, parallels remain parallel and the middle of a segment remains in the middle of the projected segment. In any other plane of perspective, the proportions are distorted. However, proportions of proportions remain unchanged, since the cross ratio is a projective invariant. What does this mean for perception? When we walk in an urban scene, for example, all buildings deform continuously, depending on the perspective, but we feel that their forms are not actually changing. This is the invariance of the cross ratio which maintains the consistency of the scene from one perspective to another.

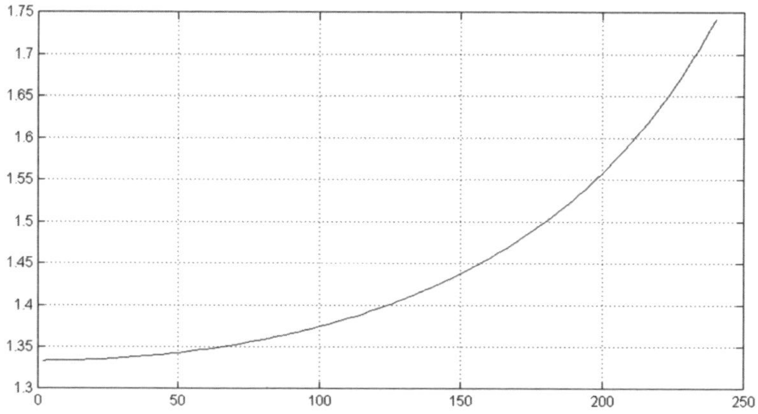

Figure 2.10. *Cross ratio as a function of the length of a circular sector divided in three equal parts*

Figure 2.11 [BEC 05] will convince us of the importance of consistence to the eye in a sequence of images. Of course, it would be better to observe an animation. However, even in this static sequence, we perceive that there is an "error", an image that does not fit, or a sudden deformation of the object. We compare, in the following illustration, the modified sequence with the "correct" sequence, which corresponds to a simple movement of the eye around a gantry that is not deformed.

We see that in the fourth image of the first sequence, the door has taken a position a little to the left: this picture is not consistent with the others, and it is something that the eye perceives immediately. We conclude, therefore, that *cross ratio is the visual figure of consistency*.

Analyzed separately, a quadrilateral can always be the central projection of a rectangle. The line connecting the intersections of opposite sides is interpreted as the vanishing line. Without loss of generality (with rotations and translations, we can bring this vanishing line onto the horizon line), we can consider that the board is perpendicular and, by construction, we deduce the principal point P and the principal distance d. The semi-circle (dotted) shows the aperture of 90°. This establishes the property of the square.

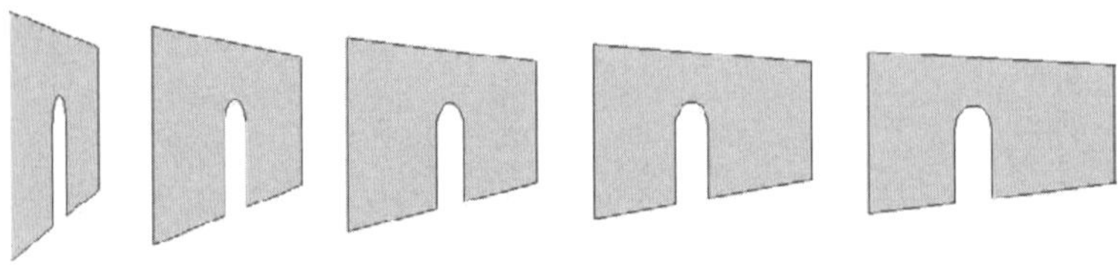

Figure 2.11. *Modification of the door position in one of the perspectives*

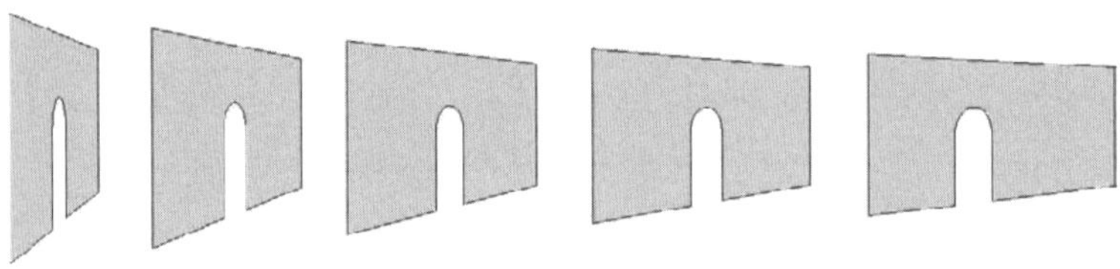

Figure 2.12. *Correct sequence corresponding to a camera moving*

From there, we get the perspective of a grid. This second construction is independent of the plane in which the square holds and is not required to locate the position of the viewpoint and the direction of the board.

Figure 2.13 shows how to compose the grid; it can be done using the diagonals of the square, but also starting directly from the property of the cross ratio. The scale must be parallel to the line joining the intersections of opposite sides of the quadrilateral (vanishing line of the plane).

The results are compared in Figure 2.14. The central projection of a square divided evenly is on the left, and the image is divided into equal intervals on the right.

The drawing on the right could be the image of a hyperbolic paraboloid like the degree 1 Coons patch built on four non-coplanar straight lines, where the intersections of opposite sides are not vanishing points [BEC 05].

Figure 2.14 shows the research by the human eyes of the simplest structure. On the left, we easily recognize a regular paving. Although distorted by perspective, the invariance of the cross ratio is respected. On the right, one could imagine an irregular paving. However, our eyes prefer to bend the plane and see a hyperbolic paraboloid with a regular paving. *Regularity* is one of the simple structures sought by looking. In the central perspective, it corresponds to the harmonic relationship.

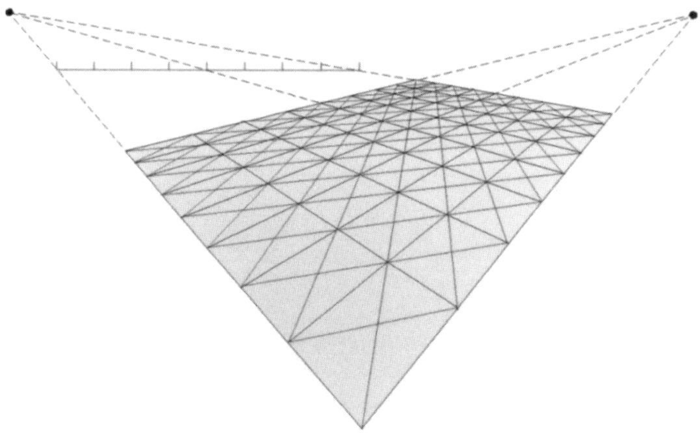

Figure 2.13. *Subdivided rectangle*

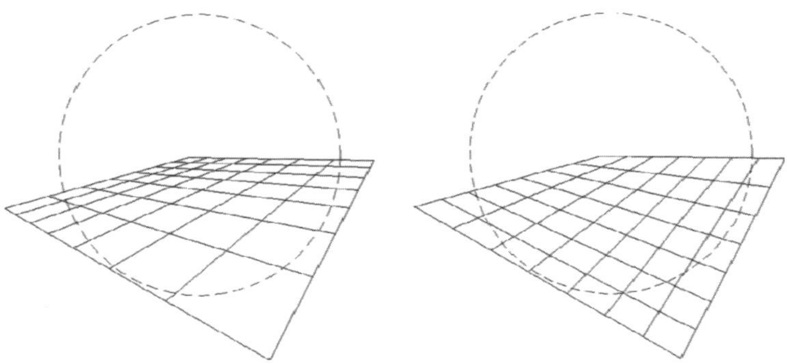

Figure 2.14. *Same objects but with angular view circle*

2.3. Harmonic relation and regularity

Let us demonstrate that the image of an *arithmetic progression* is a *harmonic progression*. The three equidistant points A, B and C are projected from a point S (origin) onto a line d (Figure 2.15). Their images are arranged according to a certain proportion, which, by homothety, persists on any parallel to the line d. It can therefore be assumed without loss of generality that the straight line passes through the point A so that the images of A, B and C are A', B' and C'. From S, let us draw a parallel to AC, which cuts $A'C'$ in P.

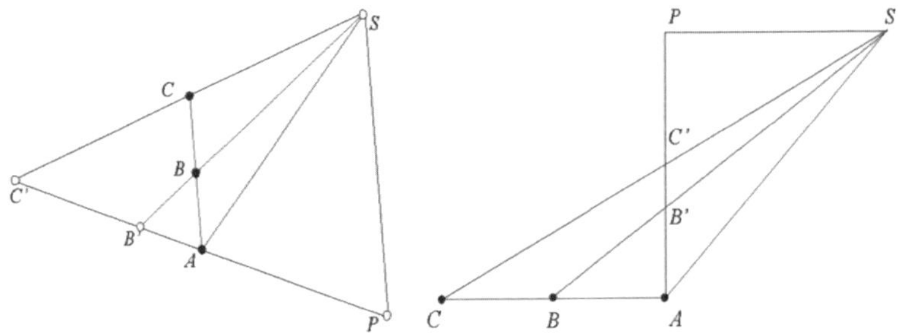

Figure 2.15. *Harmonic progression on the line [BEC 05]*

To demonstrate that the central projection transforms any arithmetic projection into a harmonic one, it suffices to show that *A*, *B* and *C* are in arithmetic progression, that is to say: if *AB* = *BC*, the distance *PB'* is a harmonic average of the distances *PA* and *PC'*. We can also argue on the two diagrams of Figure 2.15.

As the triangles *C'PS* and *C'AC* are similar:

C'P/C'A = PS/AC

so are the triangles *B'PS* and *B'AB*.

B'A/B'P = AB/PS

Multiplying both members in the previous relations, we obtain:

$$\frac{C'P}{C'A}\frac{B'A}{B'P} = \frac{AB}{AC} = \frac{1}{2} \quad [2.12]$$

This relation transforms into:

B'P. C'A = 2 B'A. C'P

B'P. AP = - B'P. C'P + 2 AP. C'P

B'P (AP + C'P) = 2 AP. C'P

to finally give the harmonic ratio:

$$\frac{1}{PB'} = \frac{1}{2}\left(\frac{1}{PC'} + \frac{1}{PA}\right) \quad [2.13]$$

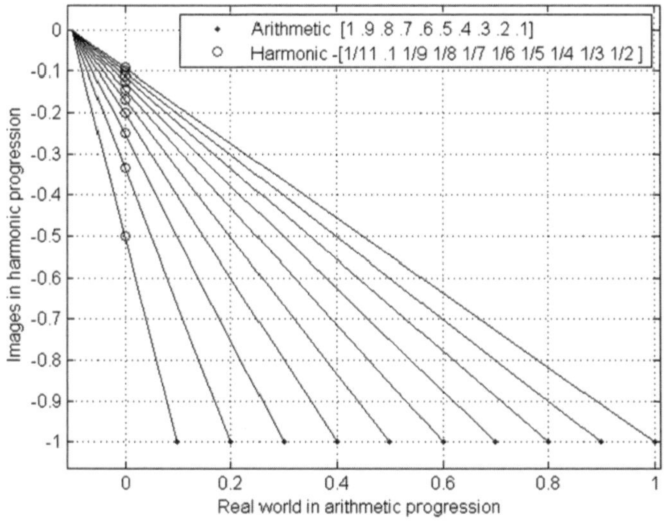

Figure 2.16. *Visualization of two progressions and their correspondence [BEC 05]*

For the AC segment, its middle B and the point at infinity (Figure (b)), the cross ratio [2.1] is:

$$\left(CAB\infty\right)=\frac{BC}{BA}\frac{\infty A}{\infty C}=-1 \qquad [2.14]$$

It has the same value as the cross ratio ($C'A\ B'\ P$):

$$\left(C'AB'P\right)=\frac{B'C'}{B'A}\frac{PA}{PC'}=-1 \qquad [2.15]$$

The cross ratio of four points in arithmetic progression (equidistant) is equal to 4/3.

To complete the comparison between the arithmetic and harmonic progressions, we take the two suites: [1. .9 .8 .7 .6 .5 .4 .3 .2 .1] and [1/11 1/10 1/9 1/8 1/7 1/6 1/5 1/4 1/3 1/2], and we arrange them, one on a horizontal axis and the other on a vertical axis (Figure 2.16). When the corresponding points are connected, the rays converge to a point that can be interpreted as the viewpoint of a central projection [PAN 75].

Conclusion: alignment, circularity, coherence and regularity.

Straight lines are a projective invariant, so alignment is guaranteed, which allows the identification of vanishing points and horizon. Conics are the visual figure of roundness, cross ratio is the visual figure of consistency and harmonic relation is the visual figure of regularity.

The examination of the invariants of the central projection on the plane establishes how our eyes perceive space, without needing to seek the recent contributions of neuroscience. In the same spirit, we now consider a more general property of the sensory world, which applies to our perception of space, but also to sound frequencies and intensities, and to light intensities. This is the foreshortening.

2.4. The foreshortening

Now we will establish the fundamental property of perception [BEC 05].

While the great thinkers of the past 2,500 years have come again and again to this statement, although they possessed, from the beginning, all the notions necessary to understand its principle and feel its scope, the fact is that nobody has formulated it clearly enough, nor have they proclaimed it in all its generality.

It probably started with the study of a stretched string, which led the Greeks (and others before them) to discover that the ratios are perceived as intervals, i.e. a geometric progression in sound frequencies becomes, for the ear, an arithmetic progression.

Apparently, the ancient philosophers never thought to generalize this law to other dimensions of perception, although they could have been aware that sound volume seems to increase regularly every time a choir doubles its staff, or the eyes feel a similar increase in lighting whenever the number of candles doubles in an enclosed space.

It is certain that the Greeks paid little attention to these energies that they could not quantify with the necessary precision. Physicists of the 17th Century developed ancient science without a methodological break and continued to think, essentially, in proportions. It was not until the industrial revolution that energies gained more importance and equations gradually replaced the old system of proportions: that was the great turning point: change a goal ("equilibrium" or "harmony") with a fully different one (the infinitesimal precision of differential geometry and the thermodynamic savings).

So, new sciences such as acoustics and lighting were born, which never could have been imagined in the vicinity of ancient thought. These new disciplines were ordered directly around the quantitative measurement. The logarithmic function, first proposed by John Napier (1550–1617), served to describe the law attributed to Gustav Theodor Fechner (1801–1887): in the sound and light intensities, a geometric progression in the stimulus results in an arithmetic progression in perception.

In acoustics, the logarithmic function was used to account for this effect, both for intensities (the *decibel*) and for frequencies (the *savart*, which is used in music as the unit of fine measuring intervals based on the logarithmic scale). In the lighting science, logarithmic filter was not imposed because economic interests (expressible in watts) always exceeded the perceptive concerns.

The frequency aspect of sound was dropped: in [STE 61], a large number of perceptive phenomena are presented, to which the Fechner law applies, almost all energetic.

This misrepresents much interpretation of the law: on energy issues, there is always a lower threshold for sensitivity, and a superior one for pain. Thus, we believe that the law allows us to be protected from the intense energies, and this is not its meaning.

The only thing that you have to protect yourself is the excess of information: the senses do not serve to investigate, but to survive; they do not count, they distinguish, and it is better to do it in the broadest possible perceptive window, but with variable sensitivity. For example: if the eye could distinguish in broad daylight with the same sensitivity as in the chiaroscuro of night, it would be canceled by the excess of information.

Thinking, like Aristotle, in terms of proportion, it seems reasonable to ask whether the visual foreshortening of distances follows the same law. But the arithmetic gives a blunt answer: no.

Indeed, in terms of visual perception of distance, we cannot say that a geometric progression in the phenomenon is perceived arithmetically.

To prove it, as throughout this chapter, we will use the central projection on the plane as a model of vision: a model in which we have learned to trust.

The cross ratio of points *A, B, C* and *D* in arithmetic progression:

$$x \quad x+p \quad x+2p \quad x+3p \qquad [2.16]$$

is written as:

$$(ABCD)=\frac{\left(\frac{CA}{CB}\right)}{\left(\frac{DA}{DB}\right)}=\frac{\left(\frac{2p}{p}\right)}{\left(\frac{3p}{2p}\right)}=\frac{4}{3} \qquad [2.17]$$

Their central projection images A', B', C' and D' are in harmonic progression:

$$\frac{1}{x} \quad \frac{1}{x+p} \quad \frac{1}{x+2p} \quad \frac{1}{x+3p} \qquad [2.18]$$

They preserve the cross ratio.

$$(A'B'C'D')=\frac{4}{3}=(ABCD) \qquad [2.19]$$

This relationship is independent of the arithmetic common ratio. We therefore deduce that four equidistant collinear points are always in cross ratio equal to 4/3 (Table 2.1).

Four points *E, F, G* and *H*, in geometric progression of common ratio *r* and located, for instance, in:

$$x \quad rx \quad r^2x \quad r^3x \qquad [2.20]$$

correspond to the cross ratio:

$$(EFGH)=\frac{(r+1)^2}{(r^2+r+1)} \qquad [2.21]$$

This time, the relationship depends on the common ratio. If we require a value equal to 4/3, then the common ratio should be equal to 1 and the four points are combined. There is therefore no alignment of points in arithmetic progression whose image is a set of points in a geometric progression, and *vice versa*.

This negative, purely geometric, demonstration was within the capabilities of the Renaissance artists and of all who think like the Pythagoreans, in terms of numbers and proportions. In fact, only the perception of sound frequency, precisely what the Pythagoreans studied, appears to be governed by a strictly logarithmic law because the harmonic structure of sound is arithmetically rigorous: its intervals correspond to exact ratios.

What happens with the Weber–Fechner law applied to the intensities? Stevens [STE 61] shows us that it is not very precise: it describes a behavior. For this reason, he uses power functions, which are more flexible and easier to calibrate.

The perception of sound intensities should be, on its physiological component, mechanical: a buffer would be the appropriate figure of the law. The eye is more complicated, due to the much smaller, quantum scale of light. There are also different visions (daytime, night and intermediate) that complicate everything.

So why not investigate the visual space in the same way, looking for a behavior law rather than an arithmetic one?

We do not seek a mathematical formula, but a perceptual property.

When the common ratio of the geometric progression tends to 0 or to infinity, the corresponding cross ratio tends to 1. It tends toward the ratio of an arithmetic progression (4/3) when the common ratio tends to 1. If the common ratio of the geometric progression is between 0.565 and its inverse 1.77, the cross ratio of the points is greater than 1.3; we can then expect this progression to look like the central projection of equidistant points.

There is an unwritten rule in architecture which states that a geometric progression, to be manageable, must have a common ratio of this order: certainly no more than 2.

2.4.1. *Variations in positions on a straight line*

Let us see three points A, B and C equally spaced on a line, and their images by central projection A', B' and C'; let us define as unit of length the distance between A' and B', and let x denote the distance between B' and C'. How far from C' is the point D' in the event that the points are in geometric progression? And in which case is the cross ratio of the images equal to that of the starting points (i.e. 4/3)?

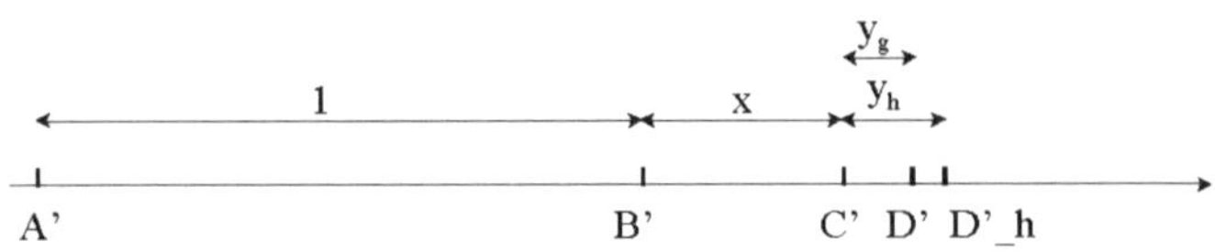

Figure 2.17. *Images on a straight line*

The common ratio of the geometric sequence A', B' and C' is:

$$\frac{B'C'}{A'B'}=\frac{x}{1}=x \qquad [2.22]$$

Therefore, if we extend the sequence to D', this point will be found at a distance of C' equal to:

$$Y_g = x^2 \qquad [2.23]$$

If the four points are located on a cross ratio 4/3, the distance C' to D' is equal to:

$$y_a = \frac{x(x+1)}{3-x} \qquad [2.24]$$

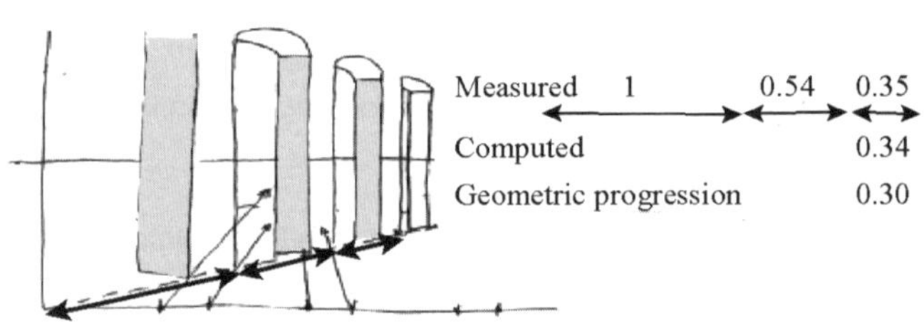

Figure 2.18. *Test of cross ratio on a freehand drawing*

We note that if $x = 3$, the corresponding harmonic progression point is located at infinity, which is easily verified by comparing the arithmetic progression [...,3, 5/2, 2, 3/2, 1, 1/2, 0] with its harmonic correspondent [...,1/3, 2/5, 1/2, 2/3, 1, 2, ∞], where the difference between 2 and 1 is three times the difference between 1 and 2/3: the following term is indeed infinite.

As shown in Figure 2.19, the interval $C'D'$ is always greater for the harmonic progression than for the geometric one. The two curves are tangent when the geometric ratio x is equal to 1.

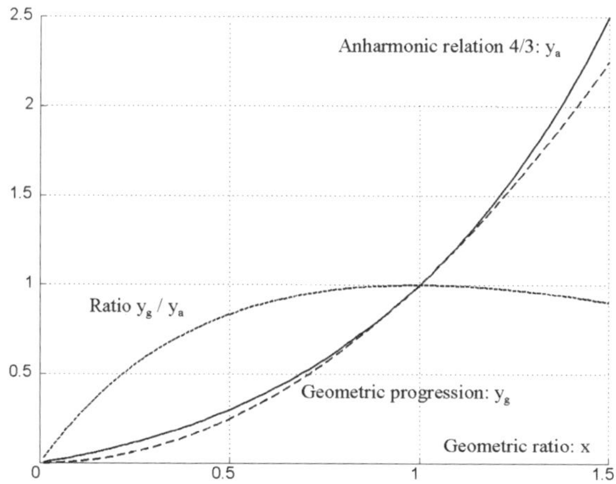

Figure 2.19. *Three transformations of geometric sequences*

These results are checked on a freehand sketch (Figure 2.18), illustrating the traditional way used to transport some perspective regular intervals (the "rule of the fourth proportional"). Despite the imprecision of the enlarged drawing, we see first that the harmonic ratio is well respected and, moreover, that the bridge piers are drawn almost in geometric progression.

2.4.2. *The critical experiment*

To be convinced of the validity of the logarithmic law in the case of light intensities, we performed the experiment proposed by Joseph Albers (1888–1976), adding layers of translucent acetate. It leaves us two clear results:

Case 1: with 1, 2, 3, 4, 5 layers (arithmetic progression), the eye perceives a decrease in range on every step, feeling a saturation effect.	
Case 2: With 1, 2, 4, 8, 16 layers (geometric progression), the eye perceives a constant interval.	

This experiment is revealed to be critical, meaning that it convinces the viewer in a simple and indisputable way that a geometric progression on the stimulus is perceived as arithmetic (case 2).

To move this experiment to the spatial dimension of the vision, we have to inhibit our habit of unconsciously interpreting the central perspective, which makes us immediately recognize the stamp of regularity in a harmonic foreshortening. As for intensities, the eye is more naive: it has not learned to interpret the energy information nor has it learnt to recognize the regularity that it actually contains in its attenuated perception of arithmetic series (case 1). Therefore, when considering the perspective, we must rely on the instructions of the rule rather than on the eye, ready to "repair" the image to recognize in it the simplest possible geometric structure.

The equivalent of the first case is shown in the figure of the bridge (Figure 2.18). Here, the equidistant piers reduce on perspective. Measuring this drawing, we have seen that the resulting harmonic progression is almost geometric, as in the case of intensities: if we extend the bridge, the intervals decrease faster and faster in the drawing, obtaining the same effect of saturation, manifested with the acetate boxes, at the horizon.

Now, to build an equivalent of the second case, we have to imagine a bridge whose arches are really in geometric progression, like the layers of acetate with a doubled number at each step. However, we do not choose a common ratio equal to 2 because we already know that it would be exaggerated. We then propose a bridge with common ratio 1.25. Figure 2.20 shows the frontal view.

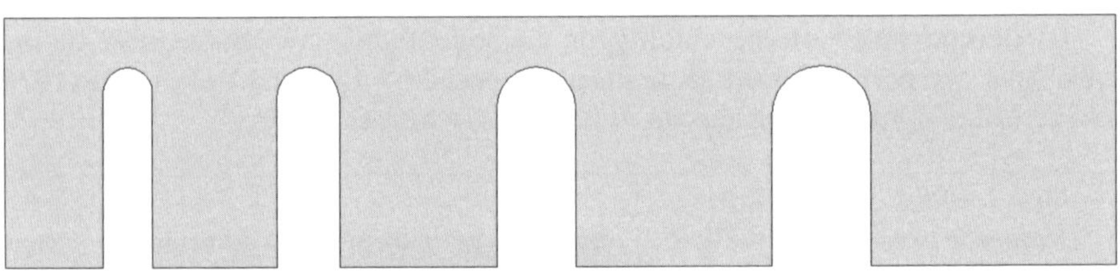

Figure 2.20. *Frontal view of the bridge with common ratio of geometric sequence equal to 1.25*

In this view, of course, the geometric progression is preserved: foreshortening is void, and also the perspective effect. Visually, this is a "degenerate case". What happens now if we look at the same bridge at an angle (30°), but from far away with binoculars?

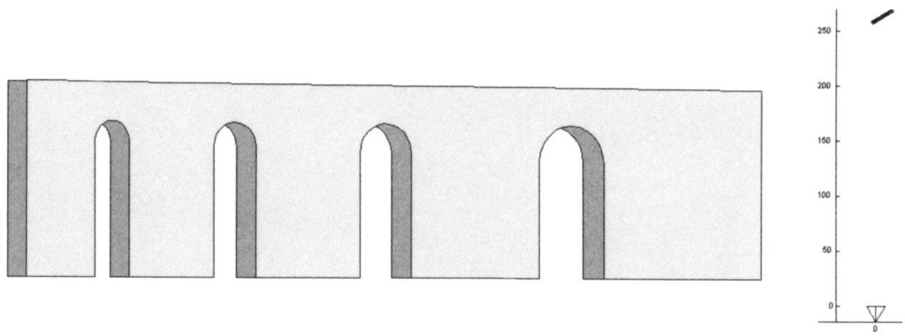

Figure 2.21. *Vision of the bridge from far away*

The geometric progression is now barely attenuated, in a somewhat bizarre sight, where the horizontals are scarcely vanishing. This time, the perspective has degenerated into axonometric, as often happens through binoculars: no foreshortening, nor perspective effect. For this reason, the reliefs are flattened when looking at distances through a lens (and also because the lens decreases the depth of field due to the optical properties of the glass).

Now, we approach the bridge, to avoid lenses, preserving the obliquity of 30°.

The horizontals vanish and the perspective effect is noticeable. The bridge is entirely within the circle of 60° of opening so that there is no deformation: the central perspective here provides a valid model for the vision. However, in the drawing, the arches are almost equidistant; the imprecision of the resulting arithmetic progression is almost indistinguishable.

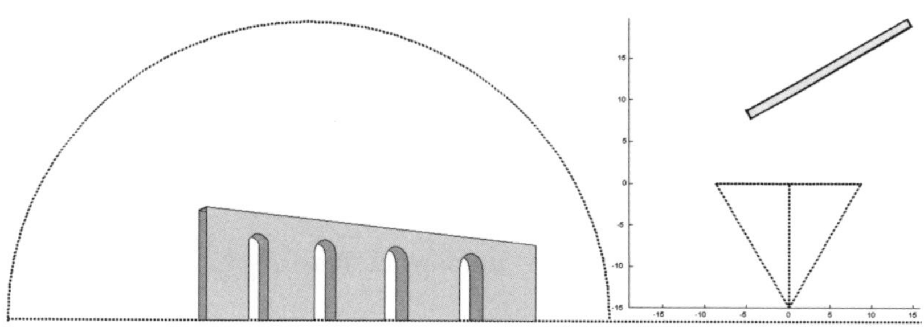

Figure 2.22. *Closer vision of the bridge*

We approach a little more now, turning the head to the right to produce an obliquity of 45°.

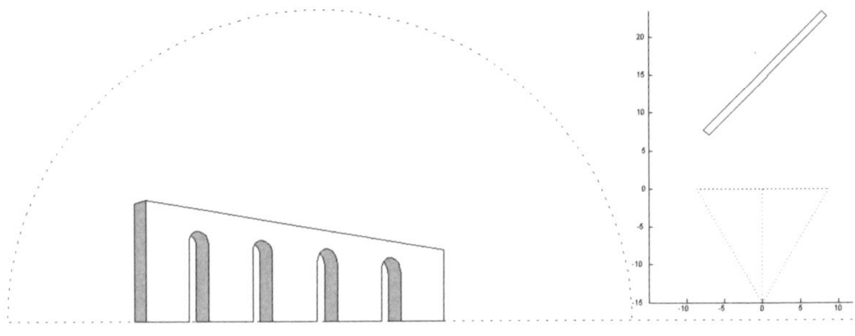

Figure 2.23. *Third position of the bridge*

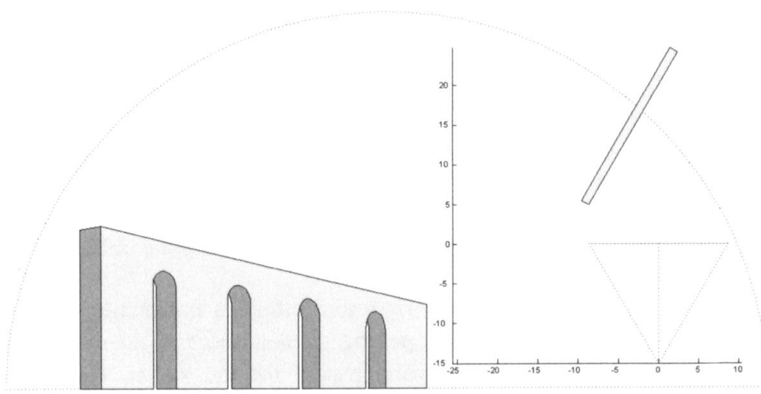

Figure 2.24. *Fourth position of the bridge*

The bridge is within the "circle of Piero", horizontals vanish more strongly and the perspective effect is much more pronounced. At the same time, the arcs are drawn very evenly spaced. We have to use a very accurate tool to ensure that the arithmetic progression rule is not perfectly accurate.

We approach a little more now, turning the head to the right to produce an obliquity of 60°.

The arithmetic progression is almost maintained in the drawing, although the case of 45° is somewhat better in this aspect.

We approach now a little more, turning the head so that our look is parallel to the axis of the bridge, creating an obliquity of 90° (principal distance $d = 15$ units).

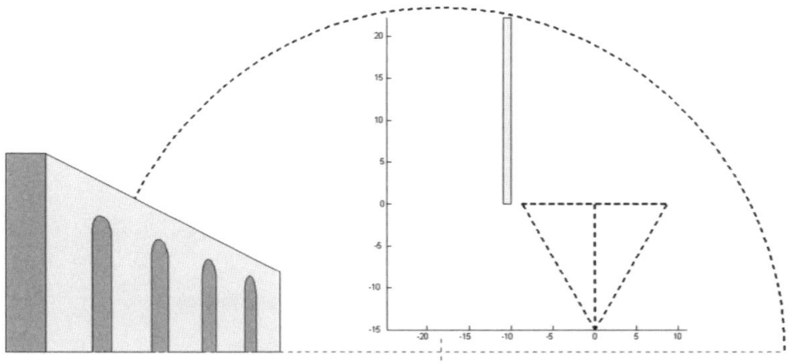

Figure 2.25. *Fifth position of the bridge*

As in the three previous configurations, drawn arches are almost equally spaced. The error is indistinguishable. So it is not a *trompe l'œil*, which would produce a particular effect in a certain position. In four very different situations, where the perspective effect is clearly visible, we note that a geometric progression in the stimulus becomes, in the perspective, an arithmetic progression.

Finally, let us look at what happens if we leave the circle of Piero, getting closer to the bridge (reducing the principal distance to the two-thirds of the previous case, i.e. $d = 10$ units).

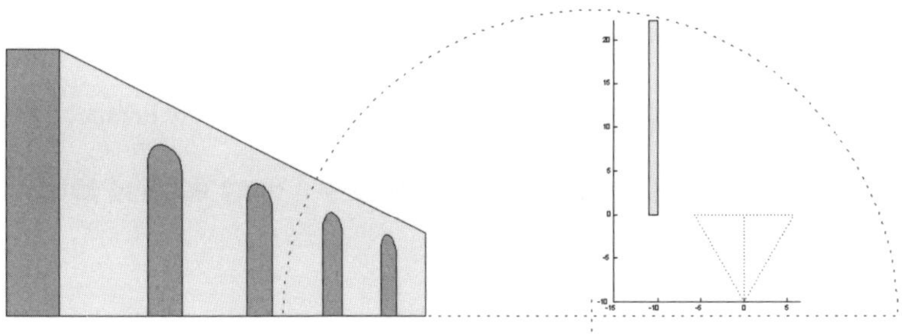

Figure 2.26. *Sixth position of the bridge*

Measuring on the drawing, we see that the arithmetic progression is lost where the opening angle exceeds 60°. However, the eye, accustomed to correct the distorted perspectives, is unable to detect the defaults. To convince us, we propose a more exaggerated case, even closer to the bridge (reducing the principal distance to a third, i.e. $d = 5$ units).

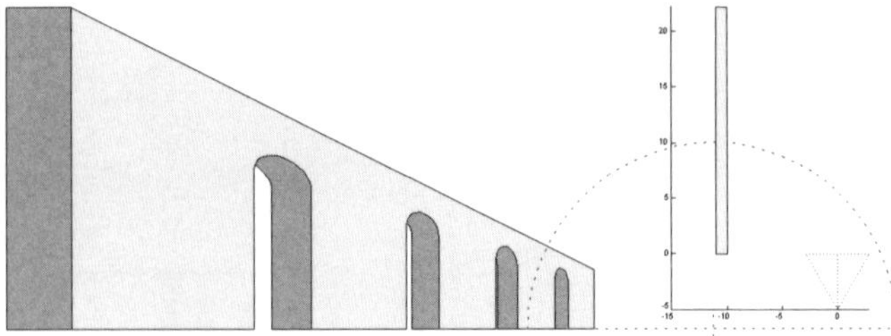

Figure 2.27. *Seventh position of the bridge*

We can conclude that a geometric progression is perceived as an arithmetic progression, provided that the following four conditions are met:

– the ratio of the geometric progression should be moderate (between approximately 0.6 and 1.7);

– the situation must not degenerate into a frontal view;

– the situation must not degenerate into an axonometric view;

– the object should not leave the opening circle of 60°.

These are precisely the same as those imposed to allow perspective effect: the object is not too deformed, the point of view does not produce a degenerate view and the opening angle is not excessive (on which case, the central projection would not provide a suitable model for the vision).

Therefore, the logarithmic operation always accompanies the perspective effect.

For a convincing conclusion, we will now give a more detailed analysis of the visual reasoning.

2.4.3. *Detailed analysis*

2.4.3.1. *The figure*

We call "GP bridge" the previously studied figure (Figure 2.20), where the cell "pile + arch" grows in geometric progression with a ratio equal to 1.25; the initial cell measures 3 units (pile of 2 units and arch of 1 unit) and evolves as follows:

Width of the four cells:		3	3.75	4.68	5.86
Start of each cell:	0	3.00	6.75	11.44	17.30

We can also build an "AP bridge" in arithmetic progression (ratio 1), starting from the same initial cell. Its total size of 14 units makes it a little smaller than the GP bridge.

2.4.3.2. *The rotation*

To facilitate schemes, we leave the observer fixed and the object rotating.

As the figure shows, the bridge is 10 units away from the frame in a frontal position. Then, after pivoting around the principal point, it reaches a position perpendicular to the board. This was the process applied in the previous examples. If the principal distance is equal to 17.3 units, the bridge is just included in the Piero cone, when it is perpendicular to the table. It is also located within the cone when oriented at 30°, 45° and 60°.

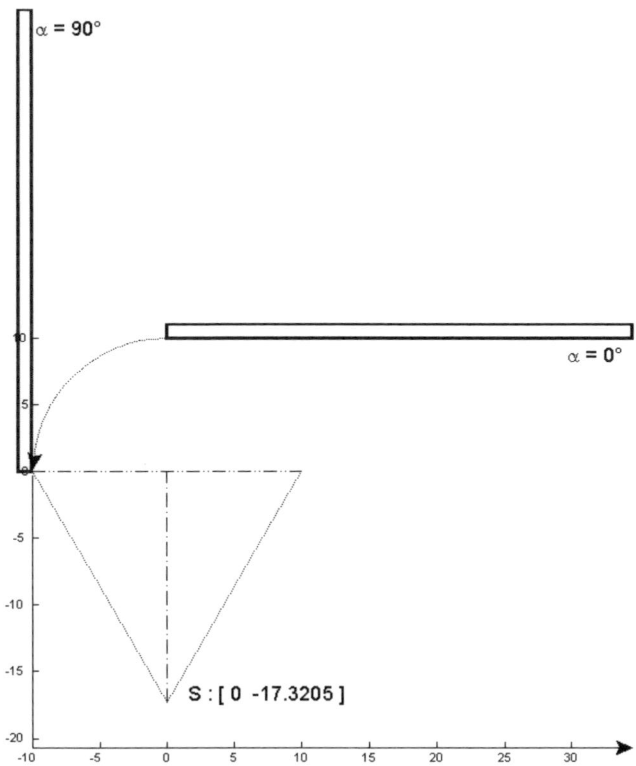

Figure 2.28. *Management of the object locations*

2.4.3.3. *The drawing*

In the frontal position, the GP bridge exhibits a geometric progression of ratio 1.25 between the sizes of the four cells in the drawing. In other directions, the geometric progression no longer seems geometric, and looks almost arithmetic. To quantify this visual observation, we have constructed three curves corresponding to the ratios between the sizes of two adjacent cells (1 and 2, 2 and 3, and 3 and 4). In the front case, the three values are equal to 1.25 and, in the case of an accurate arithmetic progression, they would be equal to 1. Overall, the three values will be slightly different, depending on the obliquity of the bridge and the distance from the observer.

In Figure 2.29, these values are represented in terms of the skew of the bridge for a principal distance of 25 units.

We consider that if the three values are between 0.9 and 1.1, the drawing looks like an arithmetic progression (the variation is almost unapparent). At a distance of 25 main units, this begins to occur from a 40° inclination, and the best result corresponds to 70°. It keeps working for 90° and then the three ratios are, respectively, 0.98, 0.96 and 0.94.

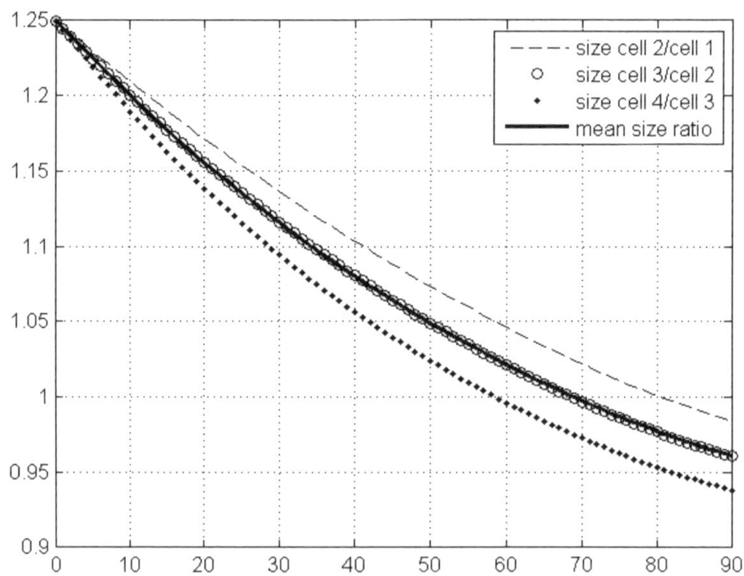

Figure 2.29. *Comparison of sizes*

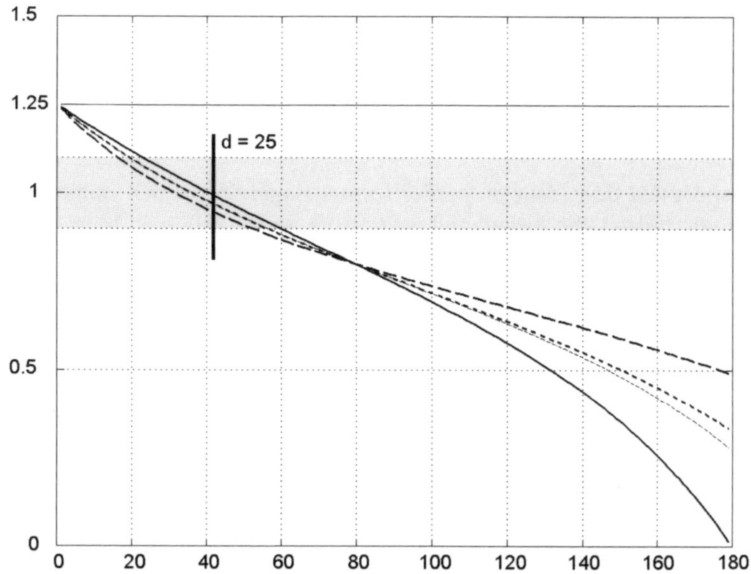

Figure 2.30. *Reference to the distance of the observer*

2.4.3.4. *The principal distance*

In addition to the obliquity of the object, the other important parameter is the distance from the observer.

In Figure 2.30, which corresponds to the bridge perpendicular to the frame, the angle including the board is varied from 0° (eye to infinity) to 180° (eye on the frame). The eye position corresponding to $d = 25$ units is indicated on the graph.

We note that the three curves come back together for $d = 11.92$ units (80° opening). In this case, the bridge is drawn with a geometric progression that is the inverse of the progression which was used to define the object.

The useful zone corresponds to an opening between 25 and 60°.

2.4.3.5. *The example*

We can now summarize the case of the GP bridge previously studied, compared to the AP bridge. We thus obtain a representation comparable to the one that we make in the case of brightness values: saturating arithmetic progression in the drawing, and a geometric progression that seems arithmetic in the drawing.

2.4.3.6. *The foreshortening*

We have just shown that the "logarithmic law" applies to the visual space as sound frequencies and intensities, both visual and audible. In these four dimensions, our perception applies a nonlinear filter, which is logarithmic.

We suspect that these four properties form a single one, which cannot be justified by physiology. Indeed, the stimuli and the sensors are too different. In its perception of frequencies, the ear adjusts to the harmonic structure of sound, arithmetically. With respect to the visual perception of space, purely of geometric nature, it essentially depends on the position of the observer, and the viewing angles. The eye perceives light quanta, while the ear repeats and damps the acoustic energies.

Only a mental process may explain similar behavior in such different contexts.

If we thought only about the energies, we might believe that this is a protective filter, but this idea makes no sense in the two other dimensions.

What do these four dimensions share, other than the extent of their perceptual windows? The ear perceives frequencies between 20 and 20,000 Hz, and intensities between 10^{-12} and tens of watts. The eye distinguishes a lit surface with fractions of lux, and supports hundreds of thousands of lux. It perceives from a few millimeters away to the stars.

In total, there is a huge amount of information, far above anything the brain can process. We should not close these windows: to survive, the human being must be able to distinguish in the daytime and at night low and high frequencies, murmurs and cries from close to far. The only solution is to compress this information in an adaptive manner (i.e. becoming adapted to the sun and the moon, noise and silence) so that the essence can always be distinguished in time.

Both nonlinear perception and relative perception are closely linked. In effect, a geometric progression may not start at zero (it cannot progress). A gray scale cannot start with absolute white or with all black. It excludes absolutes: black and white are two poles, two horizons for the scale, well-defined, but unattainable. The vision can embrace everything but its origin and infinity, its two poles. Nor is it possible to hear silence. This rejection of the poles comes from the logarithmic filter and makes relativity a prerequisite for a quick perception, better prepared to discriminate than to calculate.

<table>
<tr><th>AP Bridge</th><th>GP Bridge</th></tr>
<tr><td></td><td></td></tr>
<tr><td>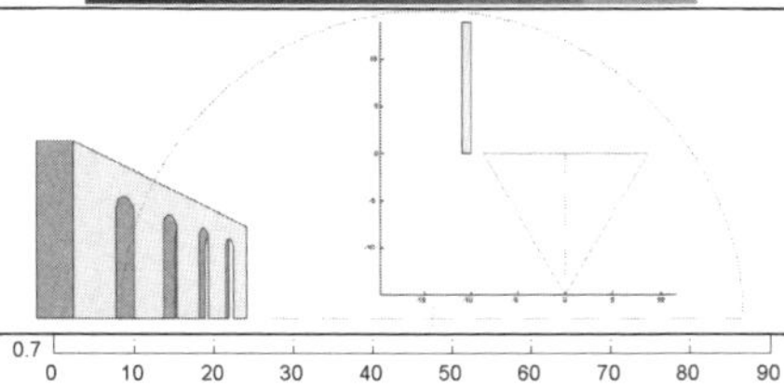
</td><td>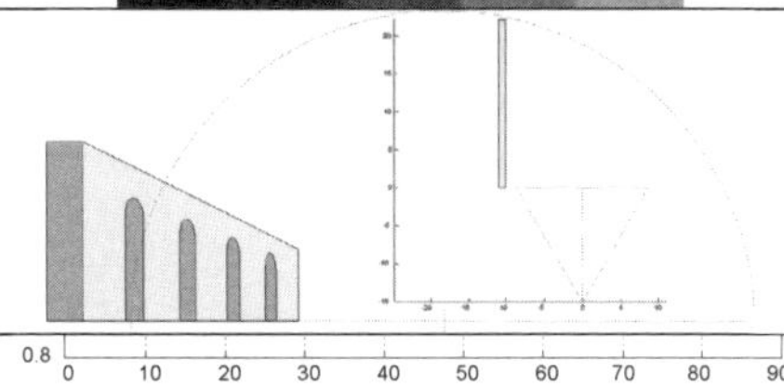
</td></tr>
<tr><td>In the AP bridge, all the cells are equal, and are projected in harmonic progression. In the chart, we see that the frontal position (zero obliquity) corresponds to a ratio 1. By increasing the inclination, the three ratios begin to diminish and separate increasingly. From about 25° inclination, the perspective effect is sensitive and the four cells no longer look the same.

In the configuration of 90°, we obtain the following values:

90°: 0.7143 0.7500 0.7778

They show the typical foreshortening of the harmonic progression. If we extend the bridge, it "saturates" on its way to the horizon, like the layers of acetate in arithmetic progression saturated on their way to the darkness.</td><td>In the GP bridge, we start above the gray area and end below it. We found that the frontal position (zero obliquity) corresponds to the preserved ratio 1.25. With increasing obliquity, we quickly enter, within 30°, into the gray area of apparent equidistance.

We obtain the following values:

00°: 1.25 1.25 1.25
30°: 1.09 1.07 1.05
45°: 1.03 1.00 0.98
60°: 0.97 0.94 0.92
90°: 0.86 0.85 0.85

In the three intermediate cases, projection displays cells nearly in arithmetic progression (as expected, the confounding case corresponds here to an obliquity of 45°), with a nearly constant step, as acetate layers in geometric progression.</td></tr>
</table>

At this point, the term "logarithmic law" begins to be uncomfortable. It is not a law, but rather a property. It is no part of arithmetic (in the modern sense of "science of numbers"), but rather sensory geometry.

It is in its application to visual space where we understand better and where we can best explain it. The first perspective specialists had a word: foreshortening (in the French of Pèlerin and du Cerceau: *raccourci*), which comes from the Italian *scorcio* (Latin *ex* and *curtiare*). In Spanish, *escorzar* is "to represent by shortening them according to the laws of perspective, things that extend in the direction perpendicular or oblique to the plane of the paper or canvas on which they are painted". It is exactly that. Piero della Francesca used the word *degradazione* which can introduce confusion. To translate the verb "to shorten", Italian uses the most common verb *accorciare*. *Scorciare* is specialized in the visual sense: it is "to shorten with the eye", i.e. visually shortening. A *scorcio* describes the concentration of information in a short space, for example, the sudden appearance of several gloomy buildings, at the opening of an alley, stacked along a perspective that opens onto the sea, there in the background, yet so close to the eye, from the direct corner giving frame to the scene.

The eye foreshortens distances and intensities. The ear foreshortens frequencies and intensities. Our brain foreshortens its perceptions and selects its information. That sounds better. Let us imagine a piano whose strings are distinguished only by their length (same width and same density). Looking at it sideways, we would see the array of strings growing in arithmetical progression: so they sound.

The big exception is color. The eye does not foreshorten frequencies, it is not needed: the window is tiny (not able to form an octave, if this term would make sense here). Therefore, from all phenomena of perception, color is the least relative.

2.5. Homogeneous coordinates

In the system of homogeneous coordinates, two-dimensional (2D)-conics and three-dimensional (3D)-quadrics are expressed as polynomials (they are actually rational polynomial functions).

We can consider the use of homogeneous coordinates as a simple method that allows us to standardize the formulation of conics and quadrics (that is why they got their name) where, as here, translations and other geometric transformations can be combined in a single structure.

Applied to the central projection on the plane, they help to reduce the calculation of the distance effect, this "visual shortening" that characterizes it, to a simple affine transformation (this terminology was introduced by Euler and extensively used by Möbius [NEE 97]). In other words, they transform the cross ratio invariant into the invariant of the axonometric projection: the conservation of proportion.

The conservation of proportion ratio becomes the conservation of proportions themselves, and we can, throughout the homogeneous coordinates, unify the formulations of all the planar projections.

The elementary object is the point. A point is not seen as a triplet of coordinates but rather in functional terms, such as a top end of a line segment.

To describe the objects, it is necessary to define a coordinate system, but it should not affect the object properties. The method must also be independent of the choice of coordinates. To do this, we must strictly distinguish between points and vectors.

The points are elements of the 3D Euclidean space E3. We denote them by italicized capital letters. More specifically, E3 is called the affine space. A point defines in it a position, for instance, the midpoint of a straight line and the gravity center of an object.

Vectors are elements of the 3D linear space R3. They are denoted by lowercase letters roofed by an arrow $\vec{u}, \vec{v}, \cdots$. When made explicit, points and vectors are represented by coordinate triplets written in line-matrix form.

$$P = [1 \quad 2 \quad 3],\ \vec{t} = [3 \quad 2 \quad 35] \qquad [2.25]$$

Although they are both expressed as triplets of real numbers, we must insist on their distinction. For any two points P and Q, there exists a unique vector pointing from Q to P. It is calculated by subtracting them component by component:

$$\vec{t} = P - Q,\ P, Q \in E^3, \vec{t} \in R^3 \qquad [2.26]$$

Moreover, for a given vector $\vec{t}$, there are an infinite number of pairs of points such that $\vec{t} = P - Q$. Indeed, if $\vec{u}$ is an arbitrary vector, $P + \vec{u},\ Q + \vec{u}$ is another pair of points that satisfies relation 2.26.

If we assign to each point $P \in E^3$ a point $P + \vec{u}$, a translation of the space is obtained. This demonstrates that the vectors are invariant with respect to translations, while points are not.

The elements of the affine space E^3 can only be subtracted to give a vector. They cannot be added because this operation is not defined for points, whereas it is for the vectors.

For points, however, there is an operation that is similar to the addition: the barycentric combination, often called affine combination. This is a weighted sum of points where the sum of weights is equal to 1:

$$X = \sum_{i=0}^{n} \alpha_i P_i;\ P_i, X \in E^3,\ \alpha_0 + \ldots + \alpha_n = 1 \qquad [2.27]$$

This expression can be rewritten as:

$$X = X_0 + \sum_{i=1}^{n} \alpha_i \overrightarrow{(P_1 - P_0)} \qquad [2.28]$$

It shows that it is a sum of a point and a vector, itself a linear combination of vectors $\overrightarrow{(P_1 - P_0)}$. The term "barycentric combination" comes from the barycenter or center of gravity.

The origin of this formulation is based on physics (the barycenter of a series of objects of masses m_i). An important special case is the convex combination, which is a barycentric combination where the coefficients α_i are all non-negative, their sum being equal to 1. The convex combination of points is always inside the points, which leads to the definition of the convex envelope of a set of points.

In 2D, if the points are represented by nails driven into a board, we obtain the convex envelope, holding a rubber band around these points. In this case, the convex hull is the smallest convex polygon including all points.

The convex hull of a set of points is characterized in that any straight line segment that connects two points of the assembly is entirely located within it. It is easy to verify that any affine transformation preserves convexity.

To define a vector from a set of points, we just have to write:

$$\vec{v} = \sum_{i=0}^{n} \beta_i P_i;\ P_i \in E^3, \vec{v} \in R^3, \beta_0 + \ldots + \beta_n = 0 \qquad [2.29]$$

It is interesting to compare expressions [2.27] and [2.29]. They differ in their result which is a point in the first case and a vector in the second case. It depends on the property of the linear combination of the points: in the first case, a *partition of unity* and in the second case a zero sum combination.

In a graphics system, the transformations used for positioning or rotating an object belong to the class of affine transformations. The basic operation defined for the points being the barycentric combination, we will base the definition of the affine transformation on the notion of barycentric combination.

A transformation Φ that transforms E^3 on itself is called affine transformation if it leaves the barycentric combinations invariant. If:

$$X = \sum_{i=0}^{n} \alpha_i P_i;\ P_i, X \in E^3 \qquad [2.30]$$

and if Φ is an affine transformation, then:

$$\Phi X = \sum_{i=0}^{n} \alpha_i \Phi P_i; \ \Phi P_i, \Phi X \in E^3 \qquad [2.31]$$

Specifically, the midpoint of a line segment has its image in the middle of the segment image, or the center of gravity of a set of points will have its image at the center of gravity of the spot images.

The affine transformation can also be written in matrix form:

$$\Phi P = P[A] + \vec{u} \qquad [2.32]$$

where $[A]$ denotes a 3 × 3 matrix, P and $\vec{u}$ being made explicit in single line matrices of dimension 3. It is easy to demonstrate the property keeping in mind the condition [2.27]:

$$\begin{aligned} \Phi \sum \alpha_i P_i &= \left(\sum \alpha_i P_i\right)[A] + \vec{u} \\ &= \sum \alpha_i P_i [A] + \sum \alpha_i \vec{u} \\ &= \sum \alpha_i \left(P_i [A] + \vec{u}\right) \\ &= \sum \alpha_i \Phi P_i \end{aligned} \qquad [2.33]$$

Examples of affine transformations are:

– identity: $\vec{u} = 0$, $[A] = [I]$, identity matrix;

– translation: $\vec{u}$ translation vector, $[A] = [I]$;

– scaling: $\vec{u} = 0$, $[A]$ is a diagonal matrix whose terms define scaling along the three axes;

– rotation: $\vec{u} = 0$, $[A]$ is a rotation matrix;

– shear: $\vec{u} = 0$, $[A]$ is a shear matrix.

An important special case of affine transformation is the Euclidean transformation, also called rigid body motion. It is characterized by the fact that the matrix $[A]$ is orthogonal, in other words $[A]^T [A] = [I]$. This transformation does not alter the angles or lengths.

We can also show that any affine transformation can be decomposed into translation, rotation, shear and scaling.

The rank of the matrix [A] has an important geometric interpretation: if its rank is 3, the operation transforms a 3D object into another 3D object. If its rank is lower, the transformation corresponds to a projection onto the plane (rank 2) or even on a straight line (rank 1). The affine transformation of E^2 to E^2 is univocally defined by a non-degenerate triangle and its image. In E^3, the affine transformation is univocally defined by a non-degenerate tetrahedron and its image.

Assuming that the observer is linked to the system of reference axes, it is possible to consider the movement of an object in two different ways:

– the observer moves around the fixed object, which is equivalent to perform a change of reference;

– the observer is fixed and the object moves. In this case, the reference system is fixed; it is set once and for all.

This rule is respected throughout the chapter.

Rotation around z-axis:

$$\left[R_z(\theta)\right]=\begin{bmatrix}\cos\theta & \sin\theta & 0 & 0\\ -\sin\theta & \cos\theta & 0 & 0\\ 0 & 0 & 1 & 0\\ 0 & 0 & 0 & 1\end{bmatrix} \quad [2.34]$$

Rotation around x-axis:

$$\left[R_x(\theta)\right]=\begin{bmatrix}1 & 0 & 0 & 0\\ 0 & \cos\theta & \sin\theta & 0\\ 0 & -\sin\theta & \cos\theta & 0\\ 0 & 0 & 0 & 1\end{bmatrix} \quad [2.35]$$

Rotation around y-axis:

$$\left[R_y(\theta)\right]=\begin{bmatrix}\cos\theta & 0 & -\sin\theta & 0\\ 0 & 1 & 0 & 0\\ \sin\theta & 0 & \cos\theta & 0\\ 0 & 0 & 0 & 1\end{bmatrix} \quad [2.36]$$

Scaling:

$$\left[S(p,q,r)\right]=\begin{bmatrix} p & 0 & 0 & 0 \\ 0 & q & 0 & 0 \\ 0 & 0 & r & 0 \\ 0 & 0 & 0 & 1 \end{bmatrix} \qquad [2.37]$$

Shear:

$$\left[C(\gamma,\eta,\lambda)\right]=\begin{bmatrix} 1 & 0 & 0 & 0 \\ \lambda & 1 & 0 & 0 \\ \gamma & \eta & 1 & 0 \\ 0 & 0 & 0 & 1 \end{bmatrix} \qquad [2.38]$$

Displacement or translation (or shear in the fourth dimension):

$$\left[D(\vec{t})\right]=\begin{bmatrix} 1 & 0 & 0 & 0 \\ 0 & 1 & 0 & 0 \\ 0 & 0 & 1 & 0 \\ u & v & z & 1 \end{bmatrix} \quad ; \ \vec{t}=\begin{bmatrix} u & v & w \end{bmatrix} \qquad [2.39]$$

Projective transformation matrix corresponding to a projection plane $z = 0$, and a viewpoint situated in [0 0 -*d*]:

$$\left[T_p\right]=\begin{bmatrix} 1 & 0 & 0 & 0 \\ 0 & 1 & 0 & 0 \\ 0 & 0 & 1 & \frac{1}{d} \\ 0 & 0 & 0 & 1 \end{bmatrix} \qquad [2.40]$$

By multiplying a point in the projective four-dimensional (4D) space by this matrix, we obtain a new position of the point in this space:

$$\begin{bmatrix} x & y & z & w \end{bmatrix}\left[T_p\right]=\begin{bmatrix} x & y & z & w \end{bmatrix}\begin{bmatrix} 1 & 0 & 0 & 0 \\ 0 & 1 & 0 & 0 \\ 0 & 0 & 1 & \frac{1}{d} \\ 0 & 0 & 0 & 1 \end{bmatrix} \qquad [2.41]$$

$$=\begin{bmatrix} x & y & z & \frac{z}{d}+w \end{bmatrix}$$

The return to Cartesian coordinates in the 3D affine space is achieved by applying an operator of central projection of 4D to the board in 3D, "the plane" $w = 1$. In brief, it is a central projection from 4D to 3D in the plane $w = 1$:

$$\begin{bmatrix} x & y & z & \frac{z}{d}+w \end{bmatrix} \Rightarrow \begin{bmatrix} \frac{xd}{z+wd} & \frac{yd}{z+wd} & \frac{zd}{z+wd} & 1 \end{bmatrix} \quad [2.42]$$

This new solid is defined in 3D space with coordinates representing the original solid deformed such that its orthogonal projection parallel to the z-axis corresponds to the central projection of the same object.

$$\begin{bmatrix} \frac{xd}{z+wd} & \frac{yd}{z+wd} & \frac{zd}{z+wd} & 1 \end{bmatrix} \Rightarrow \begin{bmatrix} \frac{xd}{z+wd} & \frac{yd}{z+wd} & \frac{zd}{z+wd} \end{bmatrix}$$
$$\Rightarrow \begin{bmatrix} \frac{xd}{z+wd} & \frac{yd}{z+wd} \end{bmatrix} \quad [2.43]$$

The third coordinate of the first two expressions of equation [2.43] is useful if it is necessary to perform a detection of visible faces or removal of hidden lines. The calculation of the central projection of an object is a simple operation as long as the board and the viewpoint are suitably chosen.

As the only important factor is the distance of the object to the center of projection, we may, without loss of generality, choose the plane $z = 1$ as the board and the origin as the center of projection. Then there is a simpler and faster solution for the central projection of a point or an object $Q = [x\ y\ z]$: just send a ray from it to the origin and compute the intersection of the ray with the plane $z = 1$.

$$\begin{bmatrix} \frac{x}{z} & \frac{y}{z} & 1 \end{bmatrix} \quad [2.44]$$

These central projections depend only on the orientation of the scene with respect to the z-axis and the position of the origin with respect to the object, which, in the oriented version of the projection, defines the half-space able to be present in the image. In the non-oriented version, the origin defines the vanishing plane that contains all the points that have their image at infinity. We could represent the complete space excluding the vanishing plane on the two sides of the same plane.

To illustrate the process, a cube is distorted by the transformation matrix [2.44], (Figure 2.31, dashed lines). When the orthogonal projection of the solid along the z-axis is performed, we obtain the desired result (bold lines).

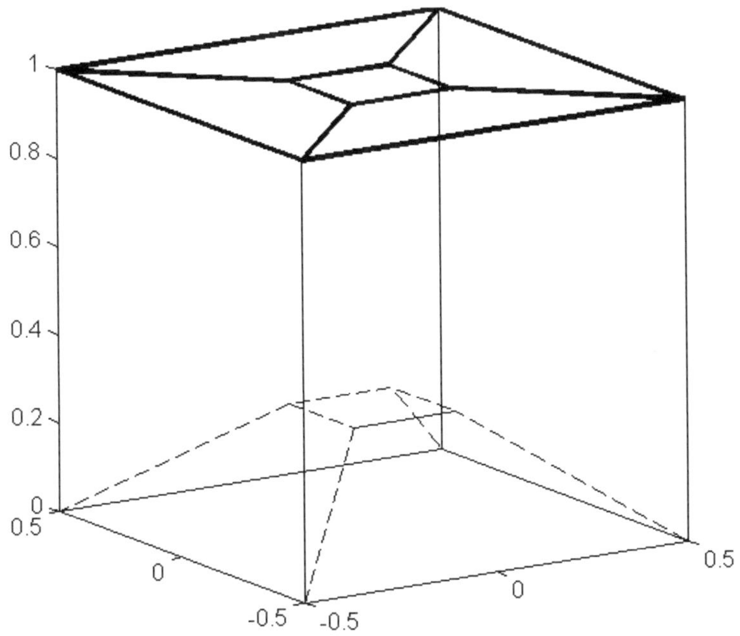

Figure 2.31. *A cube, its projective deformation and its central projection*

3

Any Scene Carried to a Sphere and the Sphere to a Point

3.1. General concepts

The central projection on the plane introduces significant distortions in the image when the opening angle is important. The images of two segments located in the same frontal plane have the same size even if their distances from the viewpoint are very different. The parameter influencing the size is the *distance to the board* not the distance to the viewpoint.

The image obtained on a flat panel is, therefore, of good quality if the opening angle is small, i.e. if the used portion of the panel is very close to the tangent point of the sphere centered on the viewpoint.

When it is necessary to consider a very large scene or to examine what happens in a large angular aperture, even in the whole 360° field surrounding the view point, a possible method is to project onto the sphere.

However, a new problem occurs because the "board" is spherical and therefore not developable. This is the problem faced by cartographers wishing to represent the Earth.

The current approach can be summarized as follows: from the scene to the sphere and from the sphere to the plane. This representation can be built "manually", but numerical simulation greatly expands the possibilities.

First, we examine the spherical object.

1) The sphere as an object: its properties, parametric (or geographical) coordinates, spherical coordinates, calculations of lengths and areas, great circles,

orthodrome and loxodrome. All the concepts and formulas of spherical trigonometry are introduced.

2) The representation of the sphere in parametric coordinates (a very common concept in the field of computer aided design (CAD)) gives the "plate carrée" and the Postel projections. The management of area conservation gives the Lambert equivalent projections. The conservation of areas also means conservation of solid angles. The extension to pseudo-cylindrical projections leads to the Hammer and Mollweide projections, widely used interstellar space representations.

3) Conservation of directions leads to the stereographic and Mercator projections. Stereography is particularly suitable to represent the sky. Finally, azimuthal representations are all reducible to one through the use of an appropriate scale [BMB 11].

Then, we examine the spherical board.

The sphere, considered as a board, allows the capture of the entire environment around the point. The interest here is to respect either the solid angle or the view factor. In the second case, we are more interested in a *hemisphere* because view factors are relative to a surface or to its tangent plane. They are seen with the right scale in the orthogonal projection of the hemisphere (Nusselt analogy).

It is also possible to work with non-spherical supports. All the azimuthal projections can be obtained from the projection on a certain surface (a kind of hemisphere), followed by an orthogonal projection from the surface to the base plane [BMB 11].

3.1.1. *Point and great circle*

In the central projection on the sphere, the board is a sphere centered at the origin coinciding with the center of projection ($O = S$). The radius of the sphere is unimportant; its value only modifies the scale of the result. Therefore, in the following, we will usually work with a unit sphere. The projection P of a point Q not located at the origin of the coordinate S, on the sphere of radius r, is given by the following formula, where $|Q|$ is the module (or distance to the center of projection) Q:

$$P = r\frac{Q}{|Q|} \qquad [3.1]$$

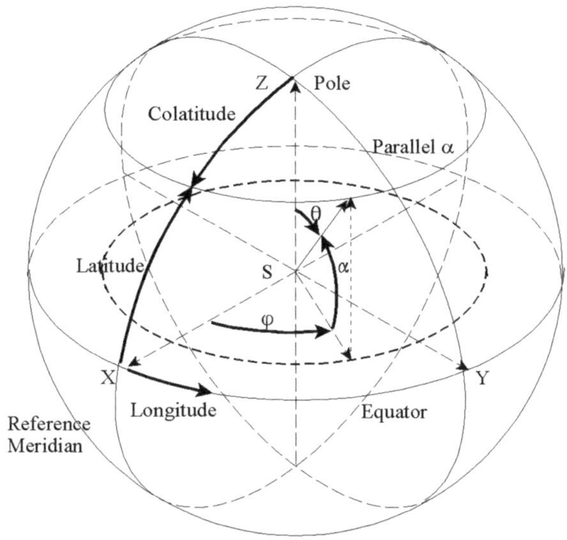

Figure 3.1. *Notations used for spherical coordinates*

After projection, the whole notion of distance from the center of projection disappears and what matters is only the direction in of the point.

This direction is marked by two angles which, in cartography, are denoted by latitude α, which varies between $-\pi/2 \leq \alpha \leq \pi/2$, and longitude φ, which varies in the range $-\pi \leq \varphi \leq \pi$. The definition of these angles requires the definition of a reference plane and therefore a great circle (the equator) and a reference point (the intersection of the equator and the Greenwich meridian). Meridians are great circles perpendicular to the equator. It finally remains to identify the two poles at the extremities of the diameter perpendicular to the equator.

In the system of spherical coordinates, *colatitude* θ (or *polar angle* or *zenith distance*) is preferred to latitude: $\theta = \pi/2 - \alpha$, with $0 \leq \theta \leq \pi$. The relationship between spherical and Cartesian coordinates is well known.

$$
\begin{aligned}
x &= r\sin\theta\cos\varphi \\
y &= r\sin\theta\sin\varphi \\
z &= r\cos\theta
\end{aligned}
\qquad [3.2]
$$

In astronomy, the position of a point on the celestial sphere is defined by the *declination* and *the right ascension*, respectively, equivalent to the latitude and longitude.

The whole space can be represented on the picture, except one point: the projection center. In the non-oriented definition of the central projection, a point has two images: the two intersections with the spherical board. In the case of a spherical board, however, keeping the two images does not make much sense because it reproduces the scene twice. By adopting an oriented definition [3.1] in the projection on the sphere, the antipodal image and the induced topological rupture are deleted.

The two images of the point define a diameter which can produce a perpendicular *great circle*. The latter is the intersection of the sphere by a plane that passes through its center. The great circle and the pair of points are *dual* to each other. Compared to this great circle, the two points are called *poles*, like in geography for equator and both geographic poles.

The great circle is marked by its inclination and the longitude of its intersections with a reference plane (the equator is a common option, but, in astronomy, the ecliptic can also be chosen, etc.). Two great circles intersecting on the reference plane, i.e. two great circles of same right ascension, may correspond to the representation of two parallel lines in space. It is observed that, on the sphere, like in the plane perspective, the distance between their images is not constant. If the difference of inclination is equal to γ, the distance between the two arcs of great circles on the unit sphere varies between γ and 0, where the images of points are at infinity. The image of the infinite line is simply half of a great circle whose ends are the two vanishing points.

3.1.2. *Line and polygon*

The intersection of a sphere and a plane is a circle whose radius is the same as that of the sphere; if and only if the plane passes through its center, it is called a *great circle*. Otherwise, its radius is less, and it is called a *small circle*, which in the limiting case of a tangent plane is reduced to a point.

Therefore, the projection on a sphere of any straight line segment is an arc of a great circle. It reduces to a point if the segment is passing through the center of the sphere (projecting line).

The image of a bundle of parallel lines is a group of arcs of great circles intersecting on a diameter parallel to the same bundle.

The image of a plane is a hemisphere, unless the plane passes through the center of projection, in which case the image is the great circle corresponding to its intersection with the sphere. Figure 3.2 shows a gridded open box and its projection on the sphere. The open face of the box is situated in a plane passing through the center of the sphere. The assembly is shown in axonometric projection without perspective effect.

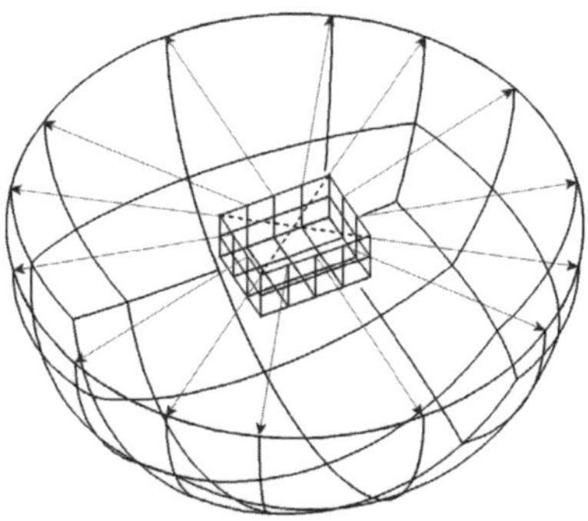

Figure 3.2. *Projection on the sphere of a $4 \times 3 \times 2$ meshed hemicube*

In the projection on the sphere, there is only one singular point, the center of the sphere. When an object has to be projected, the position of the projection center may be arbitrarily chosen, provided that it does not coincide with a point of the object.

The projection on the sphere, however, brings the problem of the flat representation, which requires a second projection and necessarily distorts the spherical image. In addition, if the picture fills the whole sphere, it is necessary to introduce an artificial line of separation in the second projection.

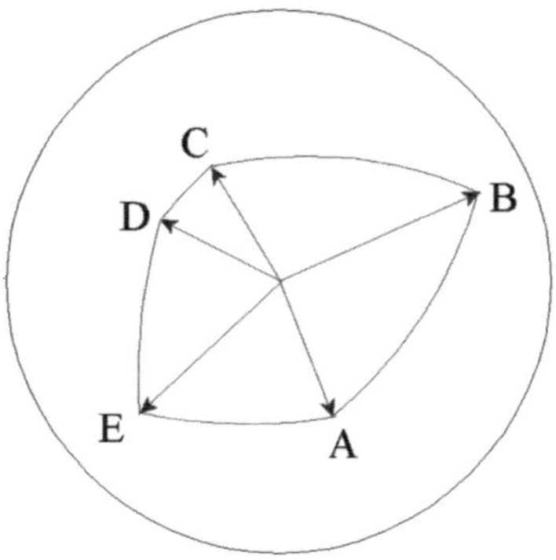

Figure 3.3. *Spherical polygon and its central polyhedron*

The image of a polyline, not necessarily flat (if it is flat, it is a polygon), is a *spherical polygon*: contour plot on the sphere composed of great circles arcs (Figure 3.3). The great circles are formed by the intersections of the sphere and planes passing through its center.

If the polyline is closed, we can distinguish:

– the sides, which are arcs of great circles of the polygon forming the contour;

– the vertices, which are the ends of the sides;

– the angles, which are the angles between two consecutive sides of the polygon.

The polyhedral central angle of a spherical polygon (Figure 3.3) is the solid angle formed by linking the vertices of the polygon to the center of the corresponding sphere. Each spherical triangle corresponds, reciprocally, to a central trihedron.

Generally, the solid angle, at a point in 3D space, is the portion of space limited by any conical surface having its vertex at that point. Polyhedral angles are particular cases. The measure of the solid angle is the portion of a spherical surface of radius equal to the unit having its center at the vertex of the conical surface limiting the angle.

Elementary meshes obtained by splitting the sphere with planes passing through its center and, therefore, based on elements of great circles			
Number of divisions	Number of elements	Characteristics of the mesh	Equivalent volume
0	1	No limit	Sphere
1	2	Monogon – hemisphere	Half-sphere
2	4	digon – spherical lune	Sphere portion
3	8	Trigon – spherical triangle	Spherical octahedron

Table 3.1. *Elementary spherical meshes built on arcs of great circles*

The elements of a central polyhedral angle are easily detected in a spherical polygon: each edge of the polyhedral angle is represented by a vertex of the polygon, each angled face by a circular arc which has the same extent and each dihedron by the curvilinear angle with the same value.

A spherical polygon is said to be convex when each extended side leaves the entire polygon in the same hemisphere; each spherical convex polygon corresponds, reciprocally, to a convex central polyhedral angle.

Unlike in the plane case, there are spherical polygons which involve only one side (half-sphere) or two sides (the lune or the sector) (Figure 3.8). Moreover, the spherical triangle is not necessarily convex, e.g. the complement of a spherical convex triangle is a non-convex spherical triangle (Table 3.1).

A spherical pyramid is the part of the sphere defined by a central angle and the corresponding spherical polygon that forms the basis of the spherical pyramid.

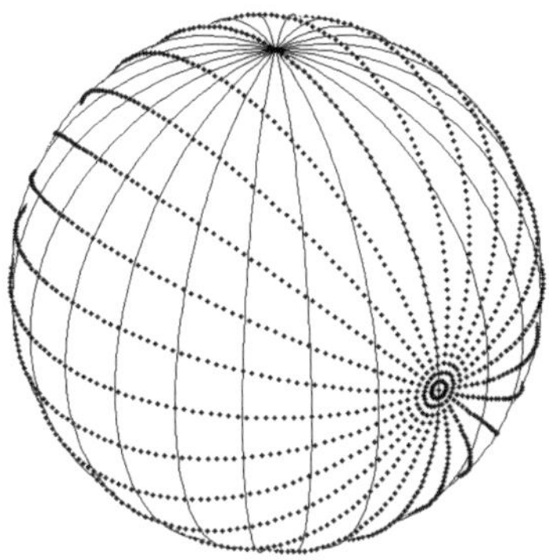

Figure 3.4. *Grid built only with great circles*

3.1.3. *Tilling of the sphere*

We limit the study to an examination of four types of meshes of the sphere where the boundaries of the elements are only circular arcs. We consider first the case of meshes created only by the arcs of great circles and therefore consisting of spherical polygons (Figure 3.4). Every great circle might represent the footprint on the earth of a satellite trajectory. The satellites follow conical trajectories included in planes passing through the center of the earth. Their traces on the terrestrial sphere are always great circles.

We will examine four possibilities.

In the first method, we consider great circles pertaining to two bundles of planes built on perpendicular diameters. Each of the great circles pivoting around an axis is characterized by its two celestial coordinates: right ascension and declination. The first angle (right ascension) indicates the position of its intersection with the equator graduated by hour meridians and measured from a reference on the equator (in astronomy, it should be the *vernal point*, where the planet crosses the ecliptic plane at the spring equinox), while the second indicates the inclination of the plane of the great circle with respect to the equator.

As can be clearly seen in Figure 3.4, the grid has the advantage of being very simple. It leads to elements that are very variable in form and very flat when approaching the plane of the two axes. Note that in the regions defined by the end of the diameter perpendicular to the two axes of rotation, the mesh is very regular and produces near square spherical figures.

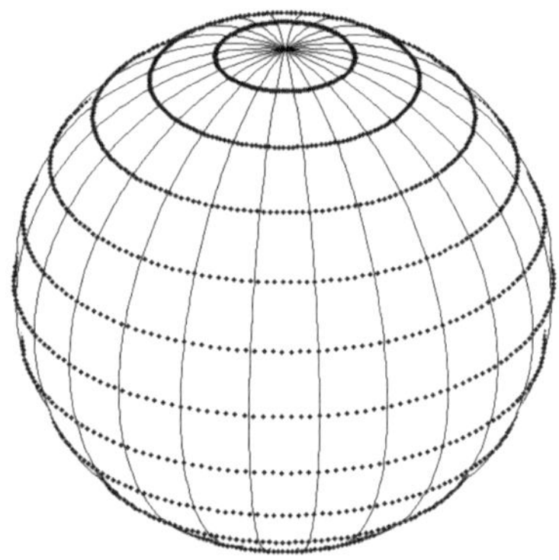

Figure 3.5. *Grid built with great and small circles*

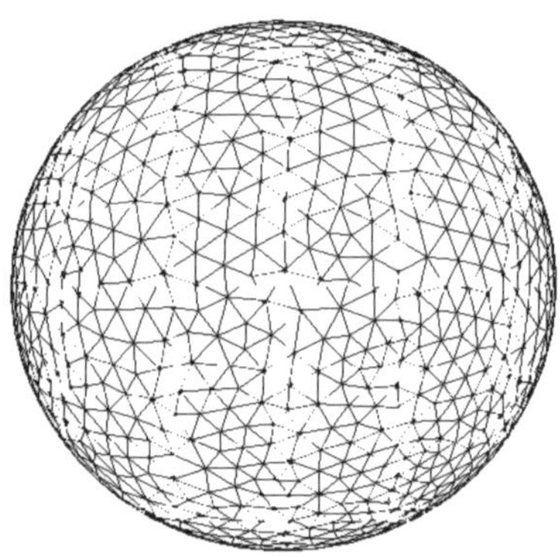

Figure 3.6. *Delaunay mesh [RAS 03]*

In the second method, the above observation leads us to use the figure of a spherical cube refined on the six faces on a regular basis as a base mesh. This method for building regular spherical polygons therefore relies on Platonic solids. It is often used, and it should be noted that among the Platonic solids, the cube and the octahedron are particularly attractive candidates. Because of their symmetry, they are naturally well adapted to the mesh of the hemisphere [SAH 98, SAH 08, WHI 98, MIG 00].

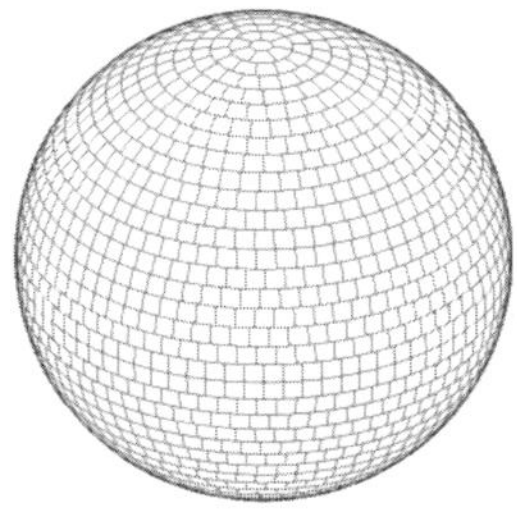

Figure 3.7. *Non congruent – 1968 cells [BEC 12a]*

A third option (Figure 3.6) is to either create a triangular mesh from the previous method and then to optimize this mesh by controlling the number of elements sharing the same vertices and the uniformity of the length of the sides with smoothing methods, or to directly create a Delaunay mesh on the sphere. These methods are well known in the field of finite elements [GEO 98, RAS 03, DU 03]; they lead to the so-called "unstructured" meshes. Their main advantage is to adapt to all forms (not only the sphere), but they require specific smoothing techniques presented and exploited by a large number of authors working in the field of finite element and structural optimization [BEC 91, ZHA 91]. Their utilization also requires the development of effective methods to navigate through the mesh.

In the fourth method [BEC 12a], we agree to use arcs of both great and small circles (Figure 3.5) to form on the sphere four-sided figures that have the advantage to be defined very simply by their spherical or geographic coordinates (latitude and longitude). This kind of tessellation (or tiling) where the covering of the entire object is achieved without an overlap or a crack is less constrained than the previous methods where vertices and edges have to be consistent (congruent meshes).

Finally, many algorithms provide explicit constructions of uniformly distributed points on the sphere. Some are based on the uniform distribution of points on a rectangle and then converted into a spherical surface by means of the inverse Lambert projection on the cylinder [SHA 96]. Another promising method [MAR 13] is based on a set of points computed on a Fibonacci spiral [SWI 06, GON 09, SAF 97, WHY 52].

3.1.4. *Areas and volumes*

The calculation of areas on the sphere is an essential step in the evaluation of solid angles. One can either take the differential elements of arcs or assess areas of spherical polygons. The area of a surface element is well defined from the differential elements.

$$\iint_S dS = \iint_S R^2 \sin\vartheta \, d\vartheta \, d\varphi \qquad [3.3]$$

This method is very effective for calculating the areas bounded by meridians and parallels cells. Indeed, for such surfaces, it is written as:

$$\int_{\theta_1}^{\theta_2} \int_{\varphi_1}^{\varphi_2} R^2 \sin\theta d\varphi d\theta = R^2 (\varphi_2 - \varphi_1)(\cos\theta_1 - \cos\theta_2) \qquad [3.4]$$

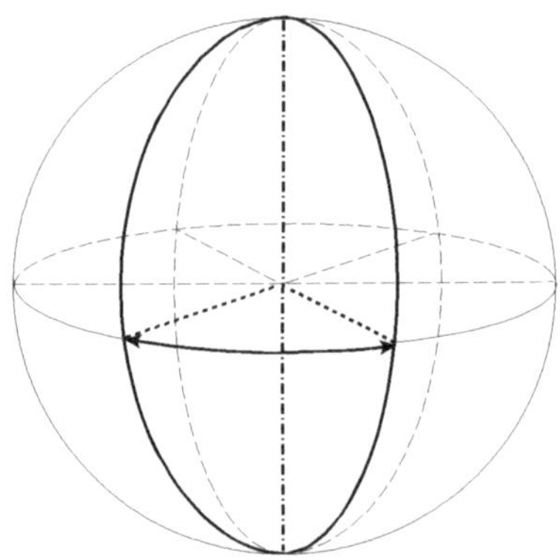

Figure 3.8. *Example of spherical lune*

From this formula, we immediately deduce the sphere area: $4\pi R^2$. A spherical cap is characterized by its extension in longitude: $\varphi_2 - \varphi_1 = 2\pi$ the position of its base $\theta_2 = \theta$ described by the opening from the pole ($\theta_1 = 0$) located at a distance h of the plane containing the limit circle:

$$\textit{Spherical cap area}: 2\pi R^2 (1 - \cos\theta) = 2\pi Rh \qquad [3.5]$$

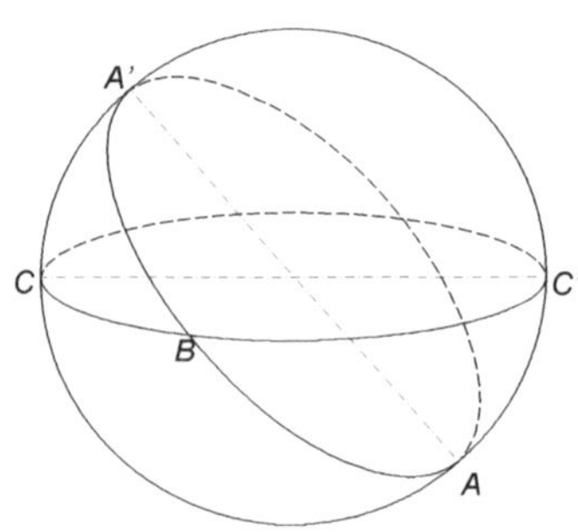

Figure 3.9. *Basic scheme for area computation*

Areas on the sphere and volumes of sphere portions can also be calculated from figures delimited by great circle arcs. We first examine the spherical lune (digon or sphere sector): the portion of the surface of a sphere which is comprised between two great semicircles ending on the same diameter (Figure 3.8). A *central dihedron* is a dihedron whose edge passes through the center of the sphere. It determines a spherical lune on the spherical surface. The spherical portion is the volume bounded by two half great circles, ending at the same diameter sphere.

When two spherical triangles of the same sphere have a common vertex and the other two are diametrically opposed, their union is equivalent to the spherical lune whose angle is the common one of the two triangles. They are called *complementary*.

In a triangle composed of the vertices of angles A, B and C, the difference $(A + B + C - \pi)$ is called the spherical excess of the triangle ABC and is represented by 2E. Note that a capital letter refers to both the vertex of a spherical polygon and the angle at that vertex, and a lowercase letter means the side of a spherical polygon and the value in radians of the arc.

When the spherical triangle is convex, we have:

$$\begin{aligned} \pi \;\leq\; A+B+C \;&\leq\; 3\pi \\ A+B-C \;&<\; \pi \end{aligned} \qquad [3.6]$$

The area of a trirectangular triangle, T, is equal to $\pi R^2/2$ (1/8 of the area of the sphere). Let ABC be a spherical triangle (Figure 3.9). Let us extend AB and CB to points A' and C' diametrically opposed to A and C; if we denote, respectively, by SL A, SL B and SL C the areas of the spherical zones that share the same angles as the triangle and by (ABC) the areas of the triangles ABC, etc., then:

$$(ABC) + (CBA') = \text{SL } A$$

$$(ABC) + (BA'C) = \text{SL } B$$

$$(ABC) + (\text{ABC}') = \text{SL } C$$

By adding, we get:

$$2(ABC) = \text{SL } A + \text{SL } B + \text{SL } C - \tfrac{1}{2}\, area_{\,sphere}$$

By dividing each term by the area of a right spherical lune whose measure is π, we obtain:

$$\frac{2(ABC)}{\pi R^2} = \frac{2A}{\pi} + \frac{2B}{\pi} + \frac{2C}{\pi} - 2 \qquad [3.7]$$

The area of the convex triangle ABC on the sphere of radius R is given by Girard's formula.

$$(ABC) = (A + B + C - \pi) R^2 = 2ER^2 \qquad [3.8]$$

The volume of the convex spherical pyramid A, B, C is equal to $(A+B+C-\pi)$ $R^3/3$; it is the product of the area of the base by one third of the radius.

The area of a spherical convex polygon with n vertices $A, B\ldots$ is given by:

$$(ABC\cdots) = (A + B + C + \cdots - (n-2)\pi) R^2, \; n \geq 2 \qquad [3.9]$$

The sum of the angles of an n-sided spherical convex polygon is greater than or equal to $(n-2)\pi$. In case of equality, the area is zero; this is the limiting case of the plane polygon.

Two equilateral spherical triangles with the same angles at the vertices are necessarily identical (same area). Unlike what happens in plane geometry, where one can subdivide an equilateral triangle, for example, into four equilateral subtriangles with an area equal to a quarter of the area of the initial triangle, this is impossible in spherical trigonometry.

	Area	Volume
Cube (edge $2R$)	$24R^2$	$8R^3$
Sphere (radius: R)	$4\pi R^2$	$\frac{4}{3}\pi R^3$
Spherical lune (radius: R, aperture: α)	$2\alpha R^2$	$\frac{2}{3}\alpha R^3$
Cylinder (radius: R; height: $2R$) 2 bases + lateral area	Total $= 6\pi R^2$ Lateral $= 4\pi R^2$	$2\pi R^3$
Cylinder sector (aperture: α; height: $2R$)	$2\alpha R^2$	αR^3
Bicone inscribed in the cylinder and sharing the same axe as the cylinder	$2\sqrt{2}\pi R^2$	$\frac{2\pi}{3}R^3$

Table 3.2. *Comparison of solids*

The main data concerning the areas and volumes of a number of solids are compared in Table 3.3. Let us note that the areas of the sphere and the circumscribed cylinder are equal [SHA 96].

3.1.5. *Spherical trigonometry*

The basic formula of the spherical geometry reflects the relationship between a vertex angle of the triangle and the three sides (Figure 3.10). Let us consider a sphere of radius equal to 1. Let us consider the spherical triangle *ABC*. Let us orthogonally project vertices *B* and *C* onto *OA* in *P* and *Q*:

$$\begin{aligned} OQ &= \cos b \quad , \quad QC = \sin b \\ OP &= \cos c \quad , \quad PB = \sin c \\ OB \cdot OC &= \cos a = (OP + PB) \cdot (OQ + QC) \\ OB \cdot OC &= OP \cdot OQ + OP \cdot QC + PB \cdot OQ + PB \cdot QC \\ \cos a &= \cos c \cos b + 0 + 0 + \sin c \sin b \cos A \end{aligned} \qquad [3.10]$$

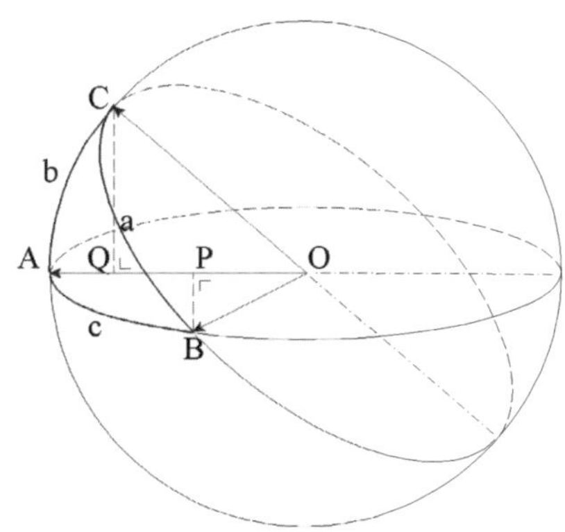

Figure 3.10. *Vertices and edges angles scheme*

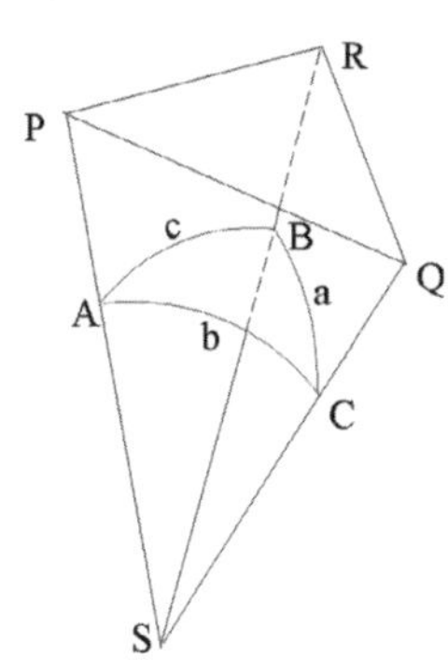

Figure 3.11. *Pyramid SPQR & spherical pyramid SABC*

The angle of the positive directions of vectors and *PB* and *QC* is equal to the angle *A*. By circular permutation:

$$\begin{aligned} \cos a &= \cos b \cos c + \sin b \sin c \cos A \\ \cos b &= \cos c \cos a + \sin c \sin a \cos B \\ \cos c &= \cos a \cos b + \sin a \sin b \cos C \end{aligned} \qquad [3.11]$$

These equations can also be solved as a function of the vertices angles:

$$\begin{aligned} \cos A &= -\cos B \cos C + \sin B \sin C \cos a \\ \cos B &= -\cos C \cos A + \sin C \sin A \cos b \\ \cos C &= -\cos A \cos B + \sin A \sin B \cos c \end{aligned} \qquad [3.12]$$

Successively eliminating the angles a and A, then b and B, and then c and C between the set of equations, we obtain:

$$\frac{\sin a}{\sin A} = \frac{\sin b}{\sin B} = \frac{\sin c}{\sin C} \qquad [3.13]$$

Finally, if we put $a + b + c = 2p$, we obtain the formula of L'Huilier [PRO 56]:

$$tg\frac{E}{2} = \sqrt{tg\frac{p}{2}tg\frac{p-a}{2}tg\frac{p-b}{2}tg\frac{p-c}{2}} \qquad [3.14]$$

This formula is useful for calculating the area of a spherical triangle from its sides.

For pyramids, these equations allow the calculation of dihedral angles of the faces from one vertex according to their angles in the same vertex (Figure 3.11). They also apply to the calculation of the dihedral angles of the faces of a polyhedron linked to a common vertex. If the vertices encompass three adjacent faces, the calculation is immediate. In other cases, it is preferable to divide the spherical polygons into triangles and analyze them separately.

Here is a simple method to calculate the angles of the spherical pyramid (Figure 3.11) from a pyramid. The projection of the triangle *PQR* on the unit sphere is the spherical triangle *ABC* where the modules of the straight lines *SA*, *SB*, and *SC* are equal to 1.

$$\overrightarrow{SA} = \frac{\overrightarrow{SP}}{\left|\overrightarrow{SP}\right|}; \; \overrightarrow{SB} = \frac{\overrightarrow{SQ}}{\left|\overrightarrow{SQ}\right|}; \; \overrightarrow{SC} = \frac{\overrightarrow{SR}}{\left|\overrightarrow{SR}\right|} \qquad [3.15]$$

We denote by A the dihedral angle of faces SAB and SAC (the others being obtained by cyclic permutation). The sides of the spherical triangle designated by lowercase letters a, b and c are arcs of great circles. Since we are on a unit sphere, the lengths of the arcs a, b and c are equal to the three angles at the vertices S of the spherical pyramid SABC; these angles are measured in radians. The perimeter of the spherical triangle is denoted as $2p = a + b + c$. The six angles A, B, C and a, b and c may be calculated from the vectors defining the pyramid.

The solid angle of the triangle PQR seen from the point of view S is equal to the area of the spherical triangle ABC built on the unit sphere with center S. To calculate the area, we proceed as follows:

Step 1: calculate the normal vectors to the faces SPQ, SQR and SRP by performing vector products between the edges of the pyramid related to the apex.

$$\overrightarrow{n_{SPQ}} = \frac{\overrightarrow{SP} \wedge \overrightarrow{SQ}}{\left|\overrightarrow{SP} \wedge \overrightarrow{SQ}\right|} \; ; \; \overrightarrow{n_{SQR}} = \frac{\overrightarrow{SQ} \wedge \overrightarrow{SR}}{\left|\overrightarrow{SQ} \wedge \overrightarrow{SR}\right|} \; ; \; \overrightarrow{n_{SRP}} = \frac{\overrightarrow{SR} \wedge \overrightarrow{SP}}{\left|\overrightarrow{SR} \wedge \overrightarrow{SP}\right|} \qquad [3.16]$$

Step 2: calculate the dihedral angles of the faces related to the apex by performing the scalar product of their normal vectors.

$$\begin{aligned} \cos A &= -\, \overrightarrow{n_{SPQ}} \cdot \overrightarrow{n_{SRP}} \\ \cos B &= -\, \overrightarrow{n_{SQR}} \cdot \overrightarrow{n_{SPQ}} \\ \cos C &= -\, \overrightarrow{n_{SRP}} \cdot \overrightarrow{n_{SQR}} \end{aligned} \qquad [3.17]$$

3.2. Cartography of the sphere

We have studied the central projection, by referring to spherical waves, because the first of the three geometric loci that define any projection is a point or, equivalently, a sphere. We then described the projection "on the sphere", this time giving a spherical shape to the second locus: the board. Cartography concerns the third locus, the object, which is spherical. This poses the problem of projecting the "sphere".

Projections "by the sphere", "on the sphere" and "of the sphere" share many characteristics, stimulate the same formulations and often raise similar difficulties.

The formulations have a general geometric range, beyond visual representation discussed later, around the notion of "solid angle".

Based on concepts developed earlier, the cartography allows us to apply different forms of previously defined projection (parallel, central and axial), studying notions and properties such as conformity or geodesics, which are fundamental, not only for the description of the land surfaces, but also in modern geometries and contemporary physics.

Compared to the previous section, the description of the sphere as an object involves introducing the concept of axis. An axis is a straight line passing through the center of the sphere, and its two intersections therewith are called poles. Of course, we immediately think of the rotation Axis of the earth, but this is only a particular case, as determined by a geographical concept that is external to geometric conditions where we place ourselves.

In this section, we will first describe in full generality the sphere with respect to an axis, and by convention we call the two intersections "North pole" and "South pole", without presuming the nature of the considered axis. The induced vocabulary naturally refers to the axis of rotation of the earth, but it can also be applied to the magnetic axis or the axis determined by any projection.

3.2.1. *Orthogonal net*

An axis determines on the sphere one single great circle perpendicular to it, which is referred to as *primary*, and an infinite number of great circles in different planes containing the axis – and thus orthogonal to that of the primary great circle – that are called *secondary*.

The primary great circle plane is called *equatorial*; the two poles of the associated axis are symmetrical with respect to this plane, at the distance of the radius of the sphere. Planes parallel to the equatorial plane determine on the sphere small-circles, *parallel*s which decrease from equator and reduce to the point of tangency at both poles.

Planes of secondary great circles are called meridians; *meridians* themselves are half great circles linking the two poles and their complements are called *antemeridians*.

The parallels and meridians form an orthogonal lattice on the sphere, since their respective tangents at the points of intersections form right angles. The parallels are not great circles, with the exception of the equator. A square canvas, formed by two parallels and two meridians, therefore, is not a spherical polygon. It is a curved rectangle with four vertices, coplanar, forming a trapezoid in plane. At the poles, tiles degenerate into bi-rectangle curved triangles, which are truncated spherical lunes.

3.2.2. *Latitude and longitude*

The equatorial plane and an arbitrary meridian plane used as a reference define a system of spherical polar coordinates to locate any point on the sphere by its latitude and longitude. The latitude of a point is the angle between the equatorial plane and the extremity of its radius: it varies from equator to the poles, between 0 and 90º. By convention, it is defined positively in the northern hemisphere and negatively in the southern hemisphere. We denote it by the letter α, and we denote by θ its complement, the colatitude measured from the associated pole.

The longitude of a point is the angle between the meridian plane passing through this point and the Greenwich meridian plane. It varies between 0 and 180º. By convention, it is taken positive east of the meridian and negative to the west. We denote it by φ. To avoid negative values, they are also known as "North latitude", "South latitude", "West longitude" and "East longitude".

A simple planar presentation of the terrestrial sphere is to construct a graph where longitudes are reported along the abscissa axis and latitudes are reported along the ordinate axis. This representation of the orthogonal network of meridians and parallels is called canvas. The diagram drawn according to the equally spaced longitude and latitude gives the so-called "plate carrée" map (Figure 3.12) which often also includes the equator, the tropics ($\alpha = \pm\ 23.5°$) and the polar circles ($\alpha = \pm\ 66.5°$).

$$\textit{Plate carrée} \rightarrow \begin{array}{l} x = \alpha \\ y = \varphi \end{array} \qquad [3.18]$$

By superimposing rectangular contours of areas of 1% of the Earth's area and by forcing most of the rectangles' widths to be equal to 2 h on the time zone graph, we can see that Antarctica is covered by more than two rectangles, i.e. more than 2% of the Earth's surface (Figure 3.13). As expected, the rectangles adjacent to the poles are highly expanding.

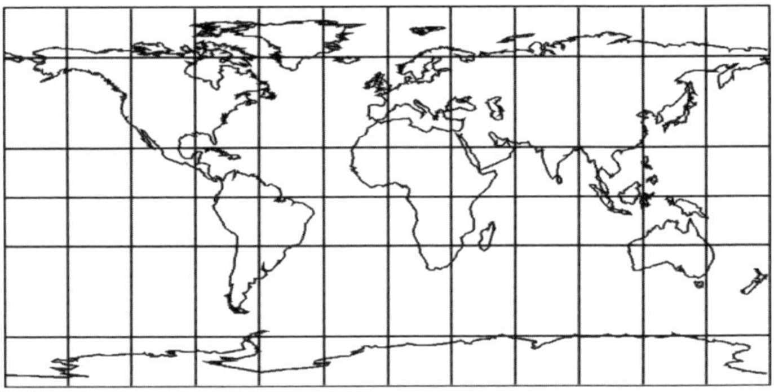

Figure 3.12. *"Plate carrée"; abscissa: longitude, ordinate: latitude*

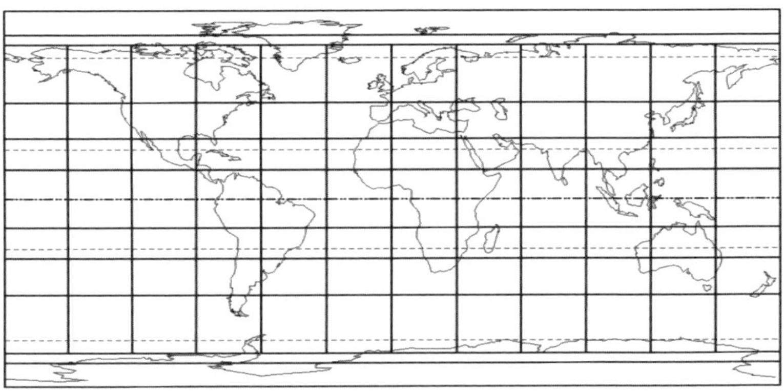

Figure 3.13. *"Plate carrée" with overlay of cells corresponding to 1% of Earth's area*

3.2.3. *Azimuth*

The *azimuth*, also called bearing, heading, course or route, is the angle between the tangent to a path defined on the sphere and the intercepted meridians, i.e. between the heading and true north, as indicated by the compass pointer (taking into account the magnetic declination).

Let us consider a great circle inclined at an angle δ (declination) on the equatorial plane that it intercepts along a diameter. Following the half great circle from the equator, the heading ς, whose initial value is the complement of the declination ($90^o-\delta$), first increases until it reaches at the vertex (the highest point of the trajectory) the value of 90°, and continues to grow to an angular value of ($90^o + \delta$) (Figure 3.14). This reasoning applies if the arcs are tracked from their right end.

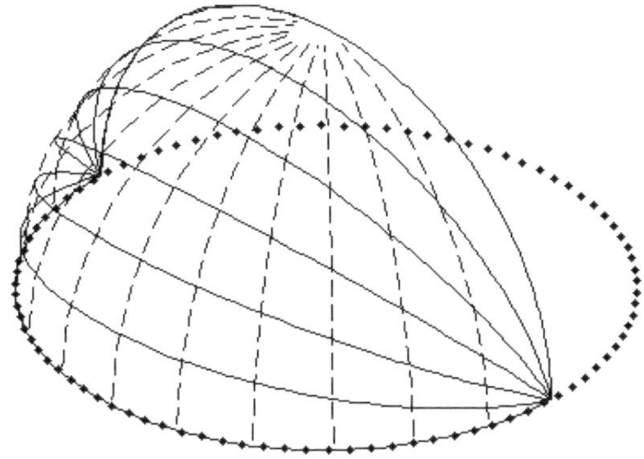

Figure 3.14. *Heading for arcs with declination from 0 to 75°*

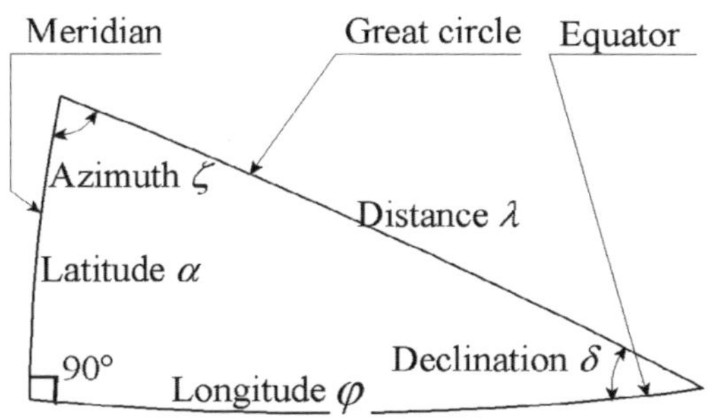

Figure 3.15. *Variables used in [3.19]*

The azimuth calculation can be done by two equivalent methods described in [3.19] and based on formulas [3.11] and [3.12]. For zero declination, i.e. when moving along the equator, the azimuth is constant and equal to 90°. We also check that, for a variation of 90°, the second formula gives a correct result, but there is only one value of the longitude and the curve of Figure 3.16 is reduced to a point. On the same figure, we have drawn the curve for a declination of 89.999°, which makes it possible to define points with longitudes ranging from 0 to 180°. The thick curve corresponds to this limit value (quite 90°). A constant heading of 90° means that the point is moving along the equator. Moving along a parallel which is not a great circle would also correspond to a constant heading of 90°.

$$\cos\varsigma = \frac{\cos\varphi - \cos\lambda\cos\alpha}{\sin\lambda\sin\alpha}; \qquad \cos\varsigma = \sin\delta\cos\varphi \qquad [3.19]$$

where:

δ orbit declination;

α latitude;

φ longitude from the right asencion;

λ arc length;

ζ heading.

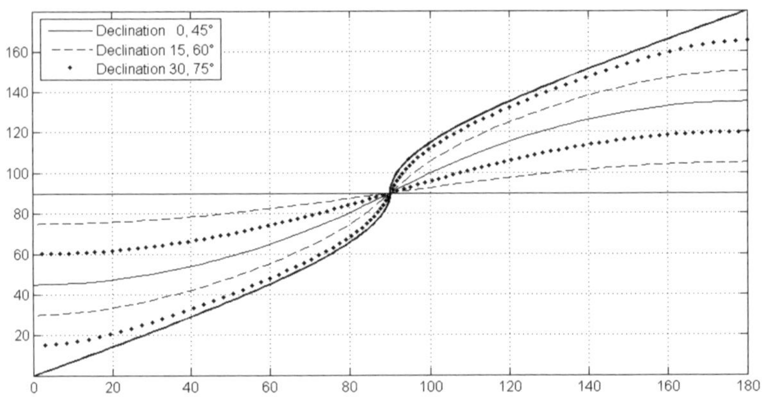

Figure 3.16. *Heading (in degrees) as a function of the position on the great circles with different declinations*

3.2.4. *Orthodromes and loxodromes*

The shortest path between two points on a surface is called *geodesic*. This is a curve without twisting, zero "geodesic curvature". The line drawn on the surface gives the impression of being right when viewed by an observer on the surface, which is why it is also called *orthodrome* (from the Greek "straight run").

On the sphere, the geodesics are great circle arcs. There are infinite geodesics joining the antipodes, but, apart from this symmetrical situation, there is only one "shorter" path joining two points of the sphere.

The arc length of a great circle linking any two points, given by their longitude and latitude (α_0, φ_0) and (α_1, φ_1), can be deduced from the fundamental formulas of spherical trigonometry [3.11]. In Figure 3.17, the lines *NP0* and *NP1* are, respectively, the meridians of longitude φ_0 and φ_1. The length of the arc is equal to the product of the angle β with the radius of the sphere.

$$\cos\beta = \sin\alpha_1 \sin\alpha_0 + \cos\alpha_1 \cos\alpha_0 \cos(\varphi_1 - \varphi_0) \qquad [3.20]$$

Another particular path is the road of constant heading, or rhumb line, or *loxodrome* (from the Greek, *oblique course*), which has the advantage of being easily determined with the help of a compass. However, only the equator and meridians are both orthodromes and loxodromes. As just shown (Figure 3.14), the path of other great circles involves a continuous variation of the azimuth: they are therefore not rhumb lines. Parallels are both small circles and loxodromes (azimuth = 90º), and therefore not orthodromes. The shortest path joining two points of the same latitude never follows the parallel, except at the equator.

Generally, the loxodrome is not a circular arc, but a logarithmic spiral (Figure 3.19). In the manual control of boats or aircraft, even in the absence of winds and currents, it is impossible to exactly follow the orthodromic, since the rhumb is constantly changing. By making regular updates, the driver simulates the orthodromic by small pieces of loxodromes.

The differential form is obtained by direct reasoning on the sphere (Figure 3.18):

$$\begin{aligned} \cos\alpha \, d\varphi &= tg\varsigma \, d\alpha \\ d\varphi &= tg\varsigma \frac{d\alpha}{\cos\alpha} \end{aligned} \qquad [3.21]$$

But, we know that:

$$\int \frac{d\alpha}{\cos\alpha} = \ln \, tg\left(\frac{\pi}{4} + \frac{\alpha}{2}\right) \qquad [3.22]$$

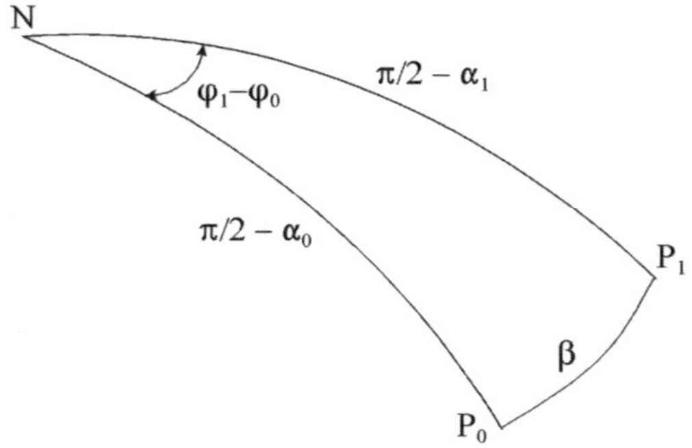

Figure 3.17. *Scheme for computing the orthodrome*

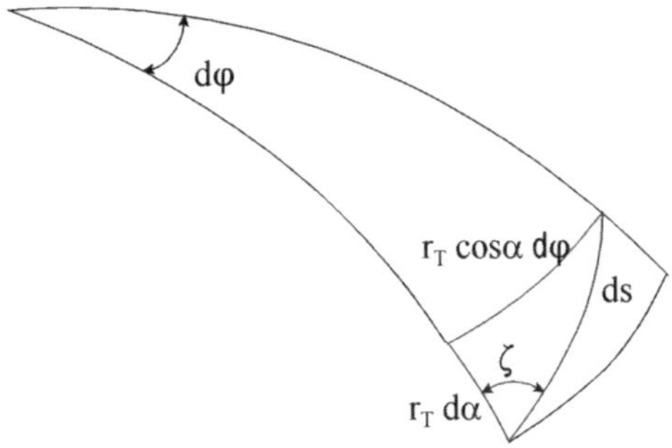

Figure 3.18. *Scheme for computing the loxodrome*

The path linking point (α_0, φ_0) to (α_1, φ_1) is characterized by the angle ς such that:

$$\varphi_1 - \varphi_0 = tg\varsigma\left[\ln\ tg\left(\frac{\pi}{4}+\frac{\alpha_1}{2}\right) - \ln\ tg\left(\frac{\pi}{4}+\frac{\alpha_0}{2}\right)\right] \qquad [3.23]$$

Because:

$$\ln tg\left(\frac{\pi}{4}+\frac{\alpha}{2}\right) = -\ln tg\left(\frac{\pi}{4}-\frac{\alpha}{2}\right) \qquad [3.24]$$

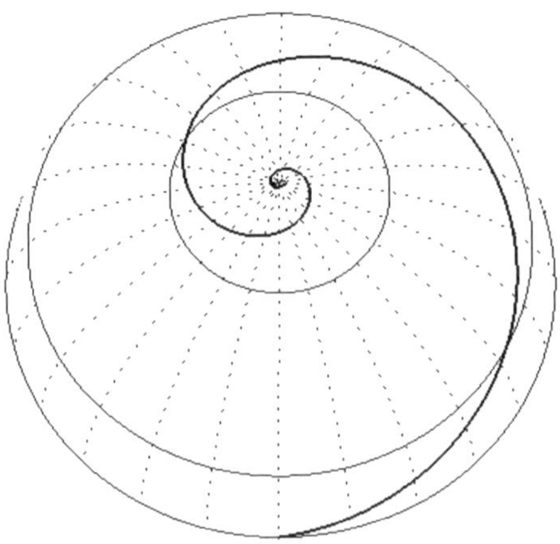

Figure 3.19. *Loxodrome (thick), hourly meridians (dotted), three reference parallels (equator, tropic and polar circle)*

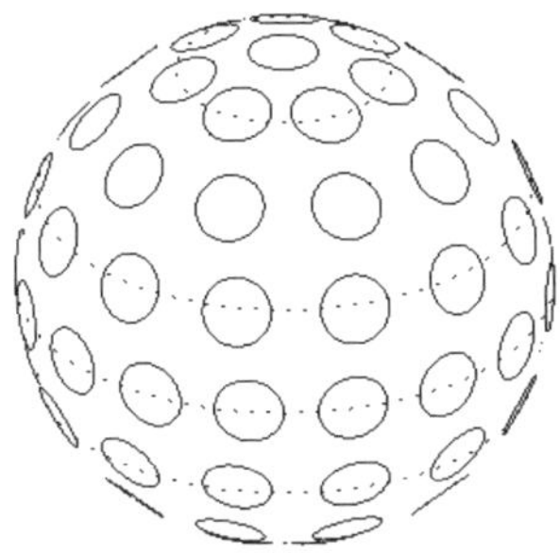

Figure 3.20. *Alterations: 78 disks on the sphere, dashed lines, parallels 0, ±23.5 and ± 66.5°*

We also obtain, by adding at the last line of [3.26] the possibility to introduce k additional rounds.

$$\begin{aligned}\varphi_1-\varphi_0 &= tg\varsigma\left[-\ln\, tg\left(\frac{\pi}{4}-\frac{\alpha_1}{2}\right)+\ln\, tg\left(\frac{\pi}{4}-\frac{\alpha_0}{2}\right)\right]\\ &= tg\varsigma\left[\ln\, tg\frac{\theta_0}{2}-\ln\, tg\frac{\theta_1}{2}\right]\end{aligned} \qquad [3.25]$$

$$tg\varsigma=\frac{\varphi_1-\varphi_0+2k\pi}{\ln\left(tg\frac{\theta_0}{2}\,\mathrm{cotg}\frac{\theta_1}{2}\right)} \qquad [3.26]$$

This is a spiral curve revolving around the pole of the sphere. Equation [3.25] can be written in a more compact form:

$$\varphi=c_0-c_1\ln\, tg\frac{\theta}{2} \qquad [3.27]$$

We will see later that this is the equation of a straight line in the Mercator projection. For two points of latitude and longitude (α_0, φ_0) and (α_1, φ_1), the length of the loxodrome (which should be multiplied by the radius of the sphere) is given by:

$$\begin{aligned}\int ds &= \int\sqrt{\cos\alpha^2(d\varphi)^2+(d\alpha)^2}\\ &= \int\sqrt{tg\varsigma^2(d\alpha)^2+(d\alpha)^2}\\ &= \int d\alpha\sqrt{1+tg\varsigma^2}\\ &= \frac{\alpha_1-\alpha_0}{\cos\varsigma}=\frac{\theta_0-\theta_1}{\cos\varsigma}\end{aligned} \qquad [3.28]$$

If the points have the same latitude, $\alpha_1 = \alpha_0$, the azimuth ς is equal to 90°. The result of formula [3.28] becomes undefined. The path is along a parallel and its length is obtained by integrating the element length. It is equal to cos α_0 ($\varphi_1 - \varphi_0$).

3.2.5. *Earth's surface shape*

The simple geometric shape nearest to terrestrial form is a spheroid flattened at the poles: the parallels are circular, but the meridian sections are elliptical.

Here, however, we limit ourselves to spherical approximation. This is the usual method to perform cartography, before making the corrections implied by a "reference ellipsoid" [SNY 93]. This operation complicates the formulations without affecting, in general, the content of the fundamental observed properties because any convex surface always projects without overlay over a sphere as long as its center is inside.

However, the ellipse and the sphere are not developable, unlike the cylinder and cone, and therefore cannot produce exact plane representation of the Earth's surface.

However, we can achieve a perspective view – indeed distorted – by projecting directly either on the plane, or on a developable surface – the cylinder or cone – which is then spread: it is the *world map*.

In this operation, we can try to minimize the error or preserve some important properties, such as lengths, angles and areas.

3.2.6. *Alterations*

A very convenient way to characterize the inevitable distortions in the flat representation of the sphere is to use the Tissot indicator [SNY 93]. The principle of the method is to draw a circle of infinitesimal radius on the sphere and compare it to its projection. The transformation takes place from plane to plane; we can demonstrate that it is affine and the representation of the circle is, in general, an ellipse. The relationship between the circle and the ellipse varies from point to point.

Instead of viewing this indicator, circles are distributed on the terrestrial sphere; we call them distortion indicators. The opening of the cones cutting the sphere is equal to $\pi/12$. The radii of these circles are equal to sin($\pi/24$), or 0.131. On the equator, we define 16 equidistant circles; at latitudes of ±22.5°: 13 circles, at ±45 degrees: 10 circles, at ±67.5 degrees: 7 circles and, finally, a circle around each pole. Note that four of these latitudes, ±22.5° and ±67.5°, are very close to the tropics (±23.5°) and polar circle parallels (±66.5°).

As shown in Figure 3.21, in the "plate carrée", the distortion is significant. The circles surrounding the poles degenerate into straight lines and the poles also appear in a straight line. Figure 3.20 also exhibits the indicators but on an orthogonal projection.

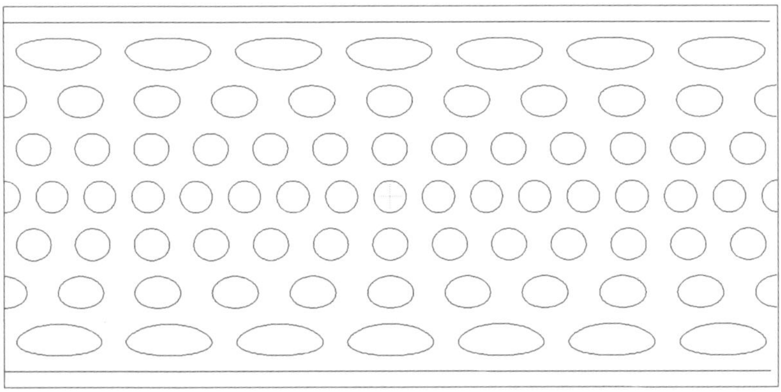

Figure 3.21. *Distortions of the "Plate carrée" shown on disks centered at 0, ±22.5, ±45, ±67.5° and ±90°*

3.2.7. *Properties of the projection*

The projection is *conformal* or *equivalent* depending on whether it preserves angles or areas. In all other cases, it is called *aphylactic*. For the sphere, the properties of conformity and equivalence are mutually exclusive.

Since all the projections introduce nonlinear changes, none can be *equidistant* in all directions. Projections are often equidistant in a singular area (e.g. the equator). Some aphylactic projections are equidistant along certain lines of the grid (e.g. meridians and parallels); they often represent a good compromise between angular alterations (lower than in the equivalent) and alterations in area (lower than in conformal). This error minimization is another possible criterion for guidance in the choice of a projection. However, the two most important and useful properties are conformity and equivalence.

3.3. Projection of the sphere on cylinders

The fundamental problem of cartography is the projection of the sphere onto a plane. It can be performed by means of a parallel, central or axial projection, on the plane, on the cylinder or on the cone; the two latter surfaces can be carried to the plane by development.

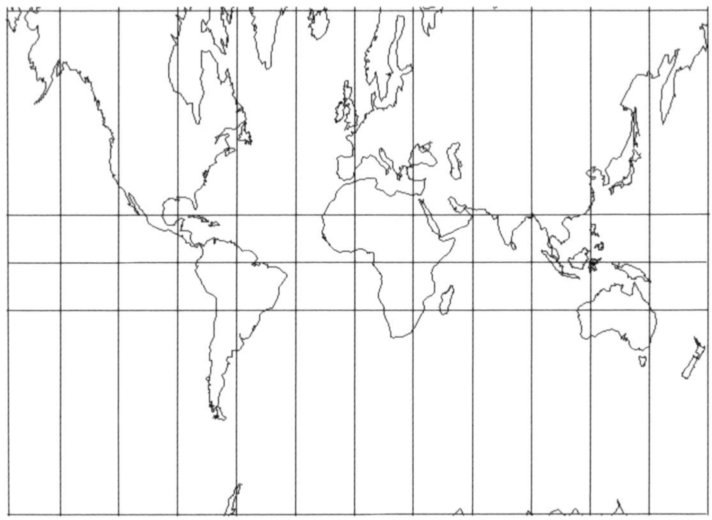

Figure 3.22. *Central projection onto the cylinder limited to the polar circles*

We do not seek the ideal projection but the respect of certain properties. When projected onto the cylinder, its axis is generally perpendicular to Ecuador. Meridians are projected along the cylinder generatrices and parallels onto circular sections of the cylinder. For a unit sphere, the length of the parallel of latitude α is equal to 2π cos α. On the sketch of the projection, it is 2π. The ratio "sketch/actual" is therefore $1/\cos \alpha$. It tends to infinity when approaching the poles.

3.3.1. *Central projection on the cylinder*

Projection rays originate from the center of the sphere. The cylinder axis passes through the center of the sphere; it may be secant or, most often, tangent to the sphere. The view obtained after the development of the cylinder is called *epicylindric* (Figure 3.22). The meridians are projected along generatrices of the cylinder and parallels by circles perpendicular to the axis. The development provides a framework, which is a grid of orthogonal lines that may involve the indications of latitude for the vertical axis and longitude for the horizontal axis.

In this projection, it is impossible to represent the poles and only Ecuador is seen in true scale. This map has very significant distortions with increasing distance from the equator (Figure 3.30); they are of the same type as in the central projection on the plane for a large opening angle.

The ratio of parallels spacing on the projection and on the sphere is:

$$\lim_{d\alpha \to 0} \frac{tg(d\alpha)}{d\alpha} \frac{1}{\cos\alpha^2} = \frac{1}{\cos\alpha^2} \quad [3.29]$$

There is little distortion between the tropics, but the ratio tends to infinity when approaching the poles.

$$\text{Cylindrical Central} \rightarrow \begin{array}{lcl} x & = & \varphi \\ y & = & tg\alpha \end{array} \quad [3.30]$$

Figure 3.23. *Lambert projection on the cylinder with aspect ratio equal to π*

3.3.2. *Lambert equal-area projection*

In this projection [LAM 1760], the sphere and the cylinder have exactly the same diameter, and the height of the outer cylinder is equal to its diameter. A simple calculation shows that the developed area of the cylinder is equal to that of the sphere. If a slice is cut from the full object (sphere more than cylinder) by means of two cutting planes perpendicular to the cylinder axis, the areas of the two sections, that of the cylinder and the sphere, are equal. In this projection, the coordinates of the map, x (along the equator) and y (along the meridians), are related to the positions on the sphere by the relation:

$$\text{Lambert on cylinder} \rightarrow \begin{array}{lcl} x & = & \varphi \\ y & = & \sin\alpha \end{array} \quad [3.31]$$

The area of the spherical cap with an opening equal to θ (meaning that the limit circle is located at colatitude θ or latitude α) has been calculated in [3.5]:

$$2\pi(1-\cos\theta) = 2\pi(1-\sin\alpha) = 2\pi h \quad [3.32]$$

The variable h is the distance from the base of the cap to its pole. On the blueprint of the Lambert projection, it is also the area of the rectangular strip between the ordinates $y = \sin\alpha$ and $y = 1$, the width of the sketch being equal to 2π.

This proves that the projection is equivalent. This property can be used to implement Monte Carlo methods with stratified sampling [SHA 96, VUE 06].

NOTE.– This projection is considered as axial because the projection rays are perpendicular to the axis of the cylinder and vary in height according to the latitude of the point to be projected. In some sense, it could also be considered as an orthogonal projection, as the projection rays are perpendicular to the cylindrical board. The ratio between the spacing of images of two parallels and their spacing on the sphere is given in [3.33]. It tends to zero when approaching the pole.

$$\frac{\sin(\alpha + d\alpha) - \sin\alpha}{d\alpha} = \cos\alpha \qquad [3.33]$$

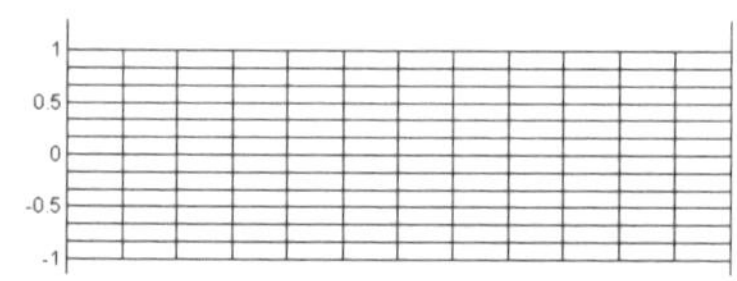

Figure 3.24. *12 × 12 rectangular uniform mesh*

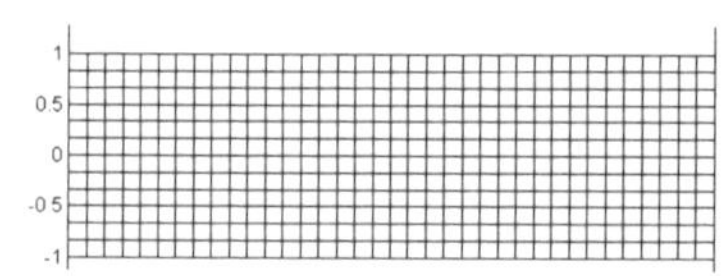

Figure 3.25. *36 × 12 rectangular uniform mesh*

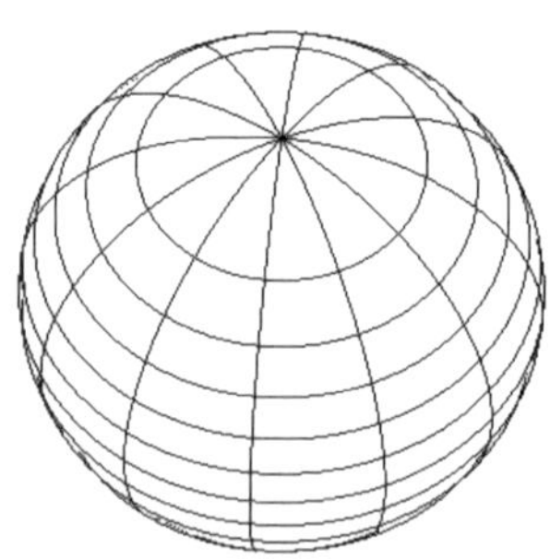

Figure 3.26. *12 × 12 equal-area cells on the sphere*

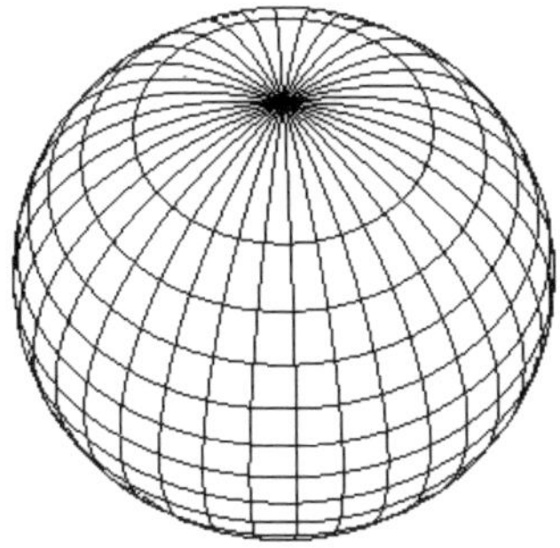

Figure 3.27. 36 × 12 equal-area cells on the sphere

3.3.3. *Mercator projection*

In the Mercator projection, we adopt the scaling factor 1/ cos α for plotting meridians, identical to the elongation of the parallels. Thus, a square delimited by two meridians and two parallels on the sphere is represented by a rectangle of aspect ratio *height*/*width* = 1/cos α.

The coordinates of the map, x (along the equator) and y (along the meridians), are related to the positions on the sphere by the relation:

$$\text{Mercator} \rightarrow \begin{array}{lcl} x & = & \varphi \\ y & = & \ln tg\left(\dfrac{\pi}{4}+\dfrac{\alpha}{2}\right)=\ln\left(tg\alpha+\sec\alpha\right) \end{array} \qquad [3.34]$$

This makes the Mercator projection a conform one. All Tissot indicators remain circular except two concentric circles at the poles, which are perpendicular to the poles' axis and therefore are represented by lines (Figure 3.29). The parts of the graph located above the upper line or below the lower one are not represented because they are extending to infinity and are located within the circles surrounding the poles. Unlike orthodromes, loxodromes appear as straight lines.

For example, in the flight Berlin–Vancouver (Figure 3.33), the shortest distance path is not shown on the straight line, but on the curve. Figure 3.28 exhibits in a dotted straight line a loxodrome which is far from being a shortest distance; it starts from Gibraltar and seems never to reach the Pole. In fact, the map extends infinitely upward, but the distances become smaller. Between two points of different latitudes, there are an infinite number of loxodromes corresponding to routes which are, to a greater or lesser extent, inclined with respect to the meridians.

Figure 3.28. *Mercator map including a loxodrome starting from Gibraltar in North-West direction*

Distortion of disks centered at 0, ± 22.5, ± 45 & ± 67.5°.

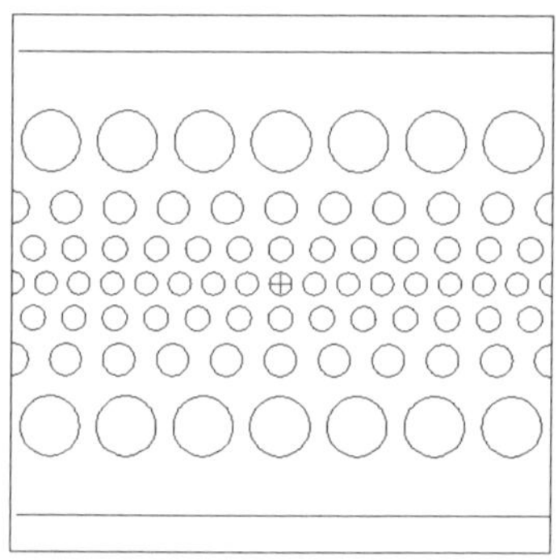

Figure 3.29. *Mercator*

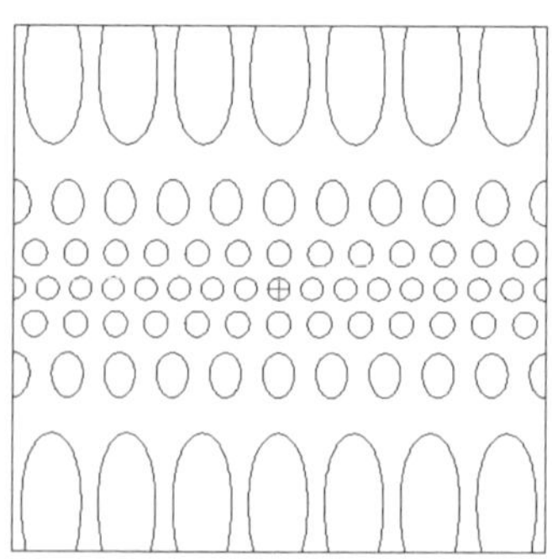

Figure 3.30. *Central on the cylinder*

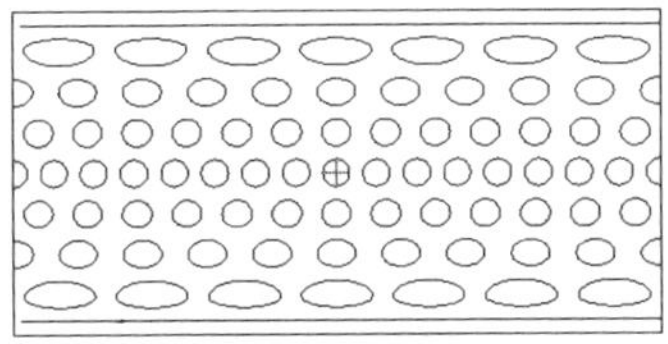

Figure 3.31. *Plate carrée*

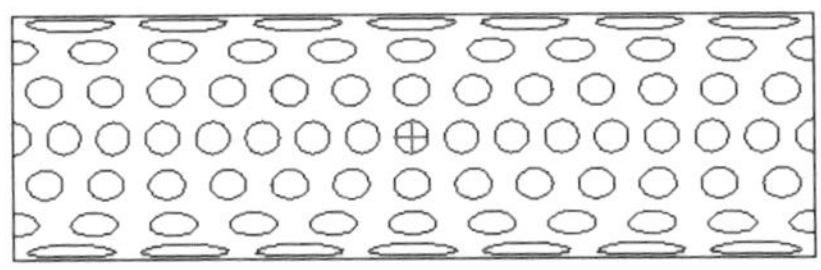

Figure 3.32. *Lambert equal area on the cylinder*

Figure 3.33. *Mercator map with orthodrome (curved line) and loxodrome (straight line) "Berlin–Vancouver"*

3.4. Projection on the plane

Projections on the plane are carried out on any plane. A projection plane can always be moved parallel to itself to make it tangent to the sphere, which only affects the scale of the drawing.

Considering that these planes are tangent to the sphere at the contact point, called the principal point of the drawing, the tangents to any curve inscribed on the sphere at these points are represented without distortion. In particular, the geographical directions or azimuths, indicating the orientations relative to the North Pole, are correctly represented at the principal point of the diagram. Another consequence of this property is that orthodromes from the principal point are always straight lines. Indeed, these are great circles passing through this point and the center of projection.

3.4.1. *Parallel projection*

A planar representation of the sphere can be obtained directly by parallel projection onto the plane. Assuming we surround the sphere with a cylinder, the projection of its outline is a circle when the projection plane is perpendicular to the cylinder axis. It is an ellipse if the plane is tilted.

The most popular configuration is the orthographic projection (Figure 3.34). Actually, the oblique projection offers little interest for the case of a sphere and is different from the previous one only by a scaling performed onto the projection. The spheres of Figure 3.39 are orthographic projections.

Figure 3.34. *Orthographic projection*

Assuming that the projection is centered on the North Pole, we obtain the relation between the spherical coordinates and the coordinates of the map:

$$Orthographic \rightarrow \begin{array}{rcl} r & = & \cos\alpha \\ \varphi & = & \varphi \end{array} \qquad [3.35]$$

3.4.2. *Central projection*

A sphere defines two specific geometric loci: its center and its surface. We can select a location of the center of projection, thereby determining the two specific central projections which are described in the following sections.

Consider first the situation where the center of projection is neither on the sphere nor in the center. If it is outside, the sphere can be surrounded by a cone which constitutes the solid angle at which it is viewed. This cone is right and its axis passes through the center of the sphere. The intersection of the cone and the board materializes the silhouette or "shadow of the sphere". If the panel is perpendicular to the axis of the cone, it is a circle. In general, it is a conical curve.

3.4.3. *Gnomonic projection*

The first interesting case is when the center of projection is at the center of the sphere. Every great circle is then projected along a straight line. This is the opposite of the projection on the sphere. Any other circle is projected by a conical curve, but the projection of a circle parallel to the projection plane remains a circle.

If the projection plane is parallel to the equator, the parallels are projected in concentric circles. Projections of meridians and parallels are orthogonal lines. In this map, we can easily find great circle air routes, actually simply connecting the two cities by a straight line. When the projection is centered on one of the poles, it is conform to the parallels and the meridians. These lines, orthogonal on the sphere, also have their images orthogonal.

The other feature of this projection is that the orthodrome, which is a great circle, is represented by a straight line, for example, on the Berlin–Vancouver route (Figure 3.35). However, the distance cannot be measured directly on the line because of the distortions.

Finally, the disadvantage of this projection is that it introduces very large deformations when approaching the edges of the hemisphere, near an opening angle of 180° (Figure 3.36); the inner circle touching the large ellipses corresponds to a latitude equal to 30°.

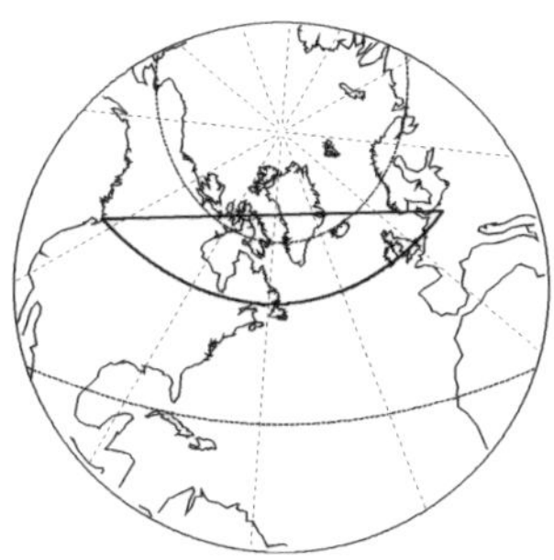

Figure 3.35. *Gnomonic including orthodrome and loxodrome Berlin–Vancouver*

Assuming that the projection is centered on the North Pole, we get the relation between the spherical coordinates and the coordinates of the map:

$$Gnomonic \rightarrow \begin{matrix} r & = & tg\theta = tg\left(\dfrac{\pi}{2} - \alpha\right) \\ \varphi & = & \varphi \end{matrix} \qquad [3.36]$$

3.4.4. *Stereographic projection*

In this case, the projection center is located on the sphere and the board is the tangent plane at its antipode. It merits special attention because it is, in fact, an *inversion*.

Figure 3.37 shows the distortion indicator corresponding to an opening greater than a hemisphere. The disks colored in gray are located on the equator. We note that the circles are not deformed, even if their size increases with distance from the main point. Circles located inside the ring formed by gray disks correspond exactly to those of Figure 3.36. Comparing this figure with the previous one, we observed that its center is less deformed and that this projection is very suitable for the representation of a half-sphere.

Figure 3.36. *Distortion in the gnomonic projection, until the 13 circles centered at latitude 22.5°*

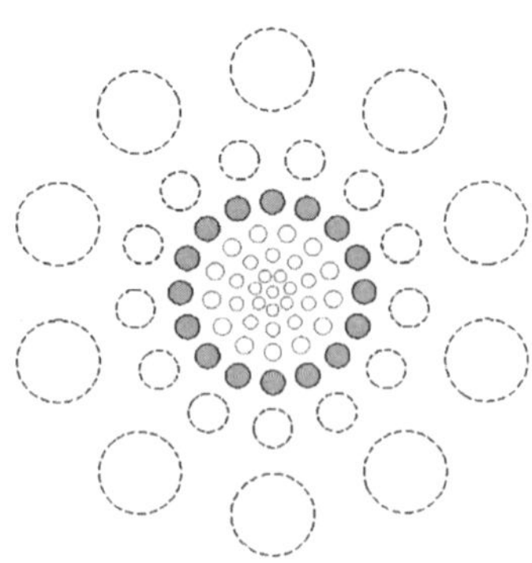

Figure 3.37. *Stereography, distortion until the limit of the external ring of circles at latitude −45°*

Assuming that the projection is centered on the North Pole, we get the relation between the spherical coordinates and the coordinates of the map located in the plane of the equator:

$$\textit{Stereography} \rightarrow \begin{array}{rcl} r & = & tg\dfrac{\theta}{2} = tg\left(\dfrac{\pi}{4} - \dfrac{\alpha}{2}\right) \\ \varphi & = & \varphi \end{array} \qquad [3.37]$$

3.4.5. *Stereography versus Mercator projection*

Let us remember the formula for computing the y coordinate in Mercator projection [3.34]:

$$y_M = \ln tg\left(\frac{\pi}{4} + \frac{\alpha}{2}\right) \qquad [3.38]$$

Using the colatitude, we obtain:

$$tg\left(\frac{\pi}{4} + \frac{\alpha}{2}\right) = tg\left(\frac{\pi}{4} + \frac{\pi}{4} - \frac{\theta}{2}\right) = tg\left(\frac{\pi}{2} - \frac{\theta}{2}\right) = \text{cotg}\left(\frac{\theta}{2}\right) \qquad [3.39]$$

and substituting in [3.38], we have:

$$y_M = \ln tg\left(\frac{\pi}{4} + \frac{\alpha}{2}\right) = \ln \text{cotg}\frac{\theta}{2} = -\ln tg\frac{\theta}{2} = -\ln tg\left(\frac{\pi}{4} - \frac{\alpha}{2}\right) \qquad [3.40]$$

We detect in [3.40] an interesting trigonometric relation:

$$\ln tg\left(\frac{\pi}{4} + \frac{\alpha}{2}\right) = -\ln tg\left(\frac{\pi}{4} - \frac{\alpha}{2}\right) \qquad [3.41]$$

Finally, we obtain a new formula for the Mercator projection:

$$\textit{Mercator} \rightarrow \begin{array}{rcl} x & = & \varphi \\ y & = & -\ln tg\left(\dfrac{\pi}{4} - \dfrac{\alpha}{2}\right) \end{array} \qquad [3.42]$$

The second line of [3.42] is the opposite of the first line of [3.37] that defines the radial component r of the stereographic projection.

3.4.6. *Postel projection*

Principles that are the basis for the representation of Postel and the "plate carrée" are the same: in both cases, the graphics are made directly by geographic coordinates. In the Postel representation, instead of using the latitudes and longitudes on a rectangular graph, we use them in a polar diagram. Longitudes are angles while colatitudes are represented by their radial arcs seen without deformation. This method allows the representation of the whole sphere, but the antipode of the principal point is seen as a circle and the distortions are very strong at the limit of the drawing (Figure 3.50).

When the projection is centered on the North Pole, the relation between the spherical coordinates and the polar coordinates on the map is:

$$Postel \rightarrow \begin{matrix} r & = & \theta = \left(\dfrac{\pi}{2} - \alpha\right) \\ \varphi & = & \varphi \end{matrix} \qquad [3.43]$$

We have noted the correspondence between the stereographic and the Mercator projections. Between the Postel and the "plate carrée" ones, there is a similar correspondence. The distances measured along radii in the first one and in vertical direction (meridians) in the second one are seen in true scale.

3.4.7. *Lambert projection*

The Lambert equivalent azimuthal projection [LAM 1760] is built as above, but the radial distance is adjusted so that each disk corresponding to a parallel on the sphere has the same area as the spherical cap delimited by the parallel. For a spherical cap ending at colatitude θ of area 2π rT 2(1-cosθ), [3.5], the radius r of the circle is:

$$\pi r^2 = 2\pi r_T^2 \left(1 - \cos\theta\right) \rightarrow r = 2r_T \sin\frac{\theta}{2} \qquad [3.44]$$

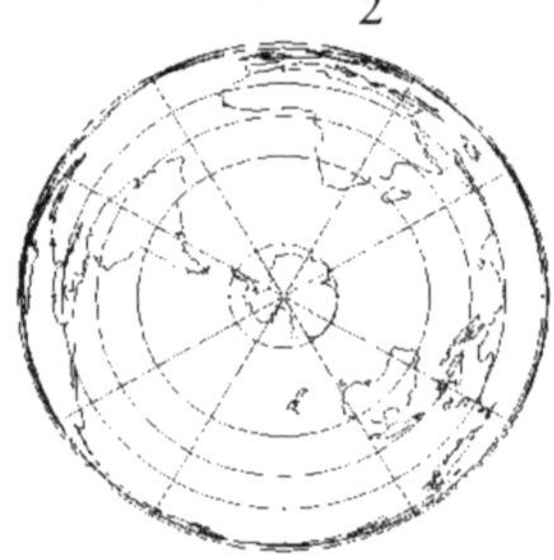

Figure 3.38. *South Pole – Lambert equal area*

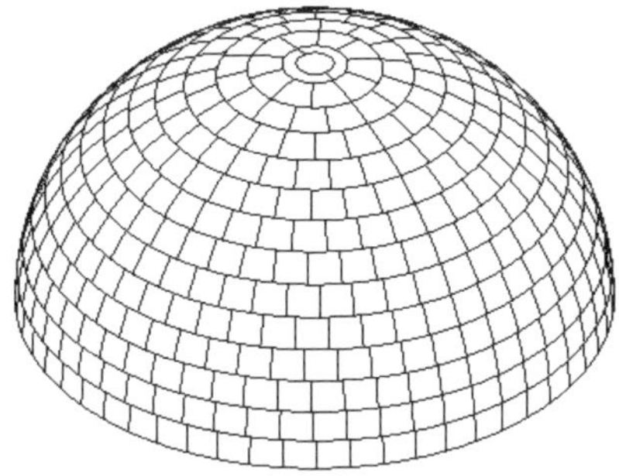

Figure 3.39. *Tilling the hemisphere with equal-area cells*

The whole spherical area is represented in a disk whose radius is equal to the diameter of the sphere. When the projection is centered on the North Pole, the relation between the spherical coordinates and the polar coordinates on the map is:

$$\text{Azimuthal Lambert} \rightarrow \begin{array}{rcl} r & = & 2\sin\dfrac{\theta}{2} = 2\sin\left(\dfrac{\pi - 2\alpha}{4}\right) \\ \varphi & = & \varphi \end{array} \qquad [3.45]$$

This sequence is directly replicated from [BBE 12]. In Table 3.3, the colatitude θ is also the angular distance defining the extension of a spherical cap centered on the pole of the hemisphere. The second column gives the radius of the disk image of the shell. The last column gives the ratio of the circumferences of the parallel limiting the spherical cap and its projection.

Colatitude θ defined on the sphere versus radial distance on the disk		
Projection	Disk radius	Spherical cap/disk circumference
Orthogonal	$\sin\theta$	1
Gnomonic	$tg\,\theta$	$\cos\theta$
Stereographic	$2\,tg\,(\theta/2)$	$(1+\cos\theta)/2$
Lambert	$2\sin(\theta/2)$	$\cos(\theta/2)$
Lamb. unit disk	$\sqrt{2}\sin\dfrac{\theta}{2}$	$\sqrt{2}\cos\dfrac{\theta}{2}$
Postel	θ	$(\sin\theta)/\theta$

Table 3.3. *Geometrical properties of the azimuthal projections*

The results of Table 3.3 are shown in Figure 3.40. We observe that the gnomonic projection is not suitable for representing the area near the equator where the colatitude is equal to 90°.

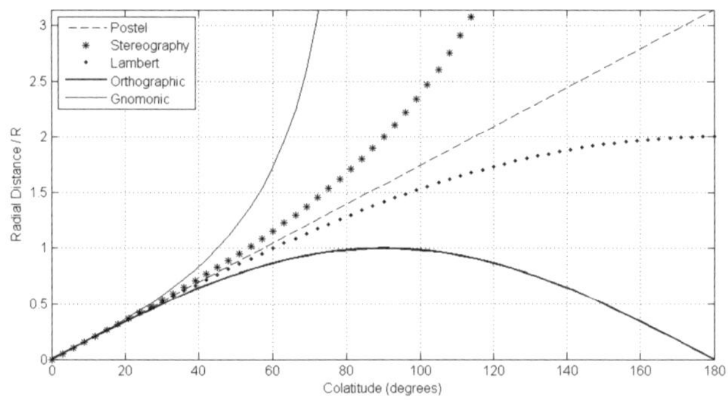

Figure 3.40. *Characteristics of five azimuthal projections [BEC 03]*

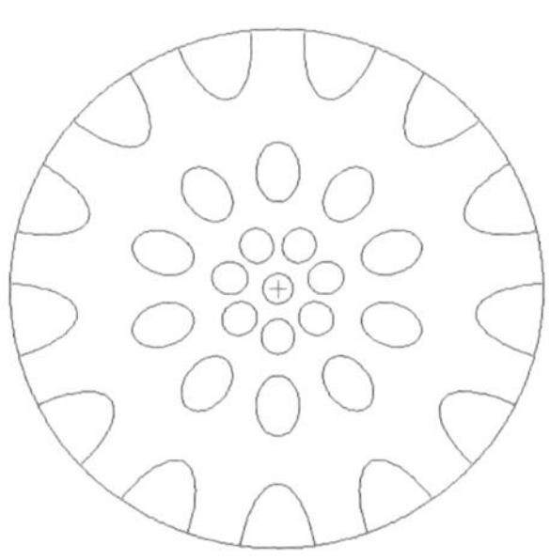

Figure 3.41. *Gnomonic*

Five azimuthal projections of the part of the sphere located at a latitude higher than the Tropic of Cancer: 23.5°– 90°

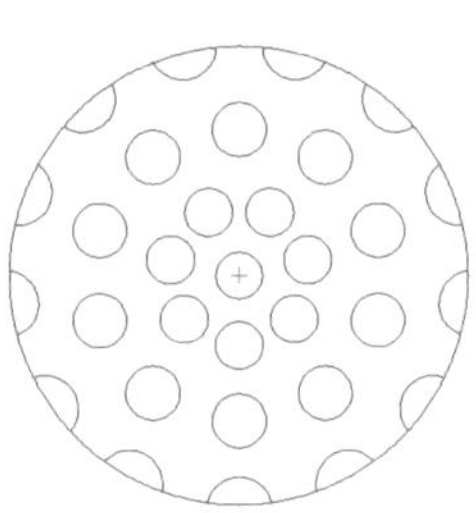

Figure 3.42. *Stereographic*

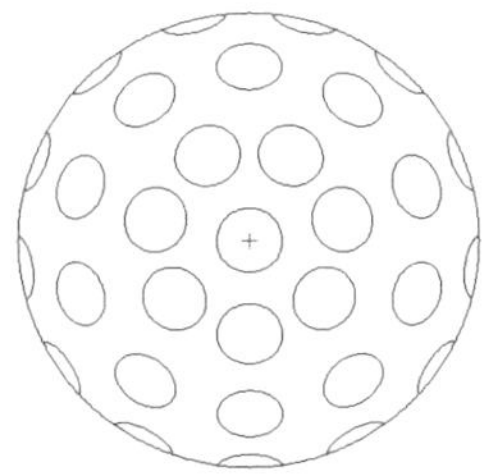

Figure 3.43. *Orthographic*

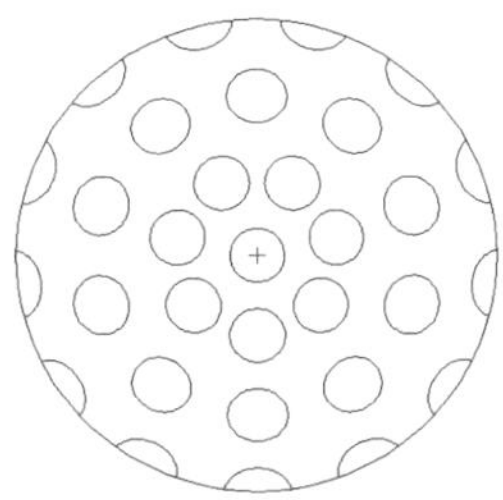

Figure 3.44. *Postel*

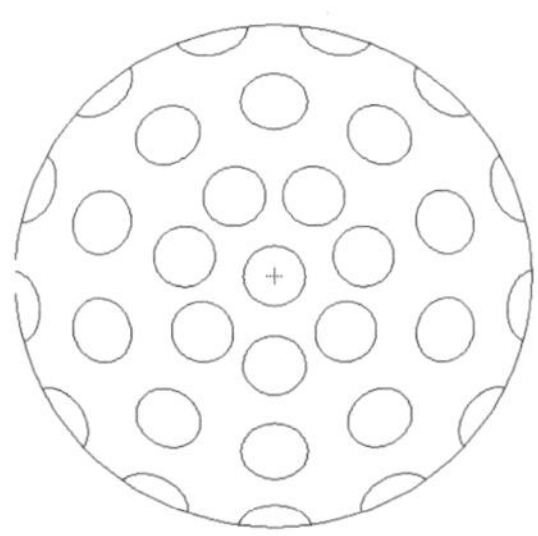

Figure 3.45. *Lambert equal area*

The orthogonal	projection	respects	the view factors
The gnomonic			the shape of orthodromies
The stereographic			the angles
The Lambert equivalent			the areas
The Postel			the distances from center

Table 3.4. *Principal characteristics of the azimuthal projections [BEC 12b]*

Among the three linear projections (gnomonic, stereographic and orthogonal), stereography is most likely to give a satisfactory representation of the areas near the equator (Figure 3.46). The main characteristics of these azimuthal projections are summarized in Table 3.4.

The distortions' indicator on the hemisphere (macroscopic version of the Tissot indicator).

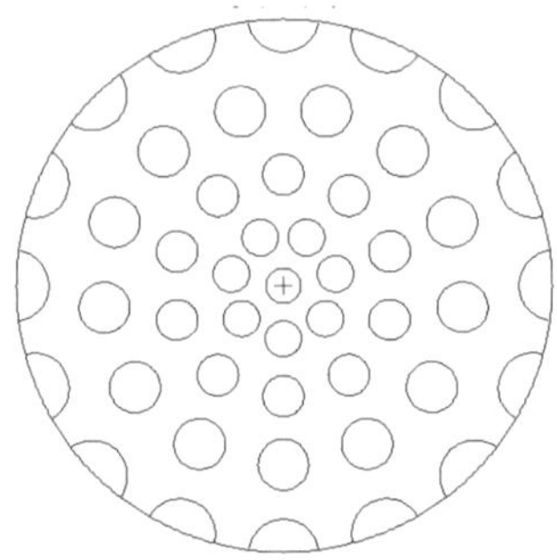

Figure 3.46. *Stereographic*

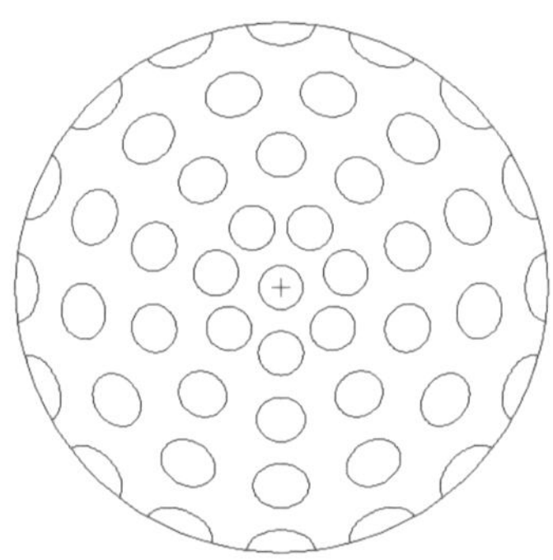

Figure 3.47. *Postel*

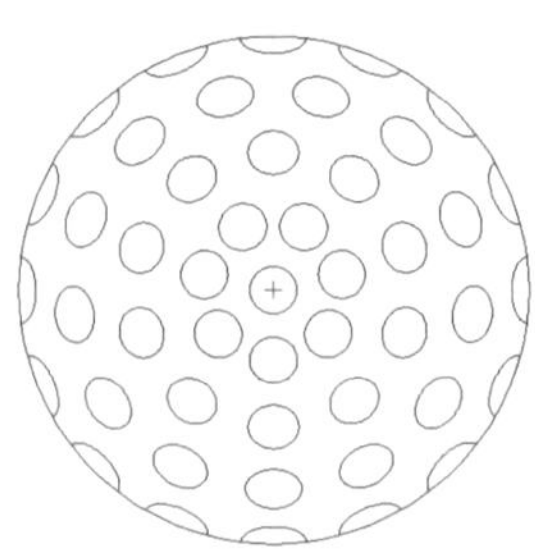

Figure 3.48. *Lambert equal area*

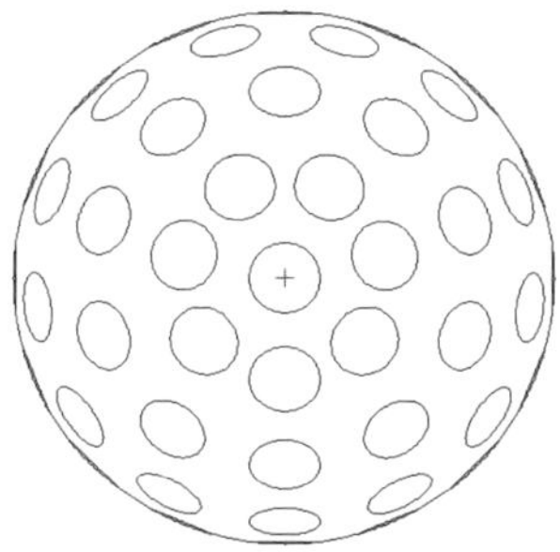

Figure 3.49. *Orthographic*

Comparison of the distortions' indicators on the sphere (macroscopic version of the Tissot indicator). Dotted lines indicate small circles passing through the disks centers

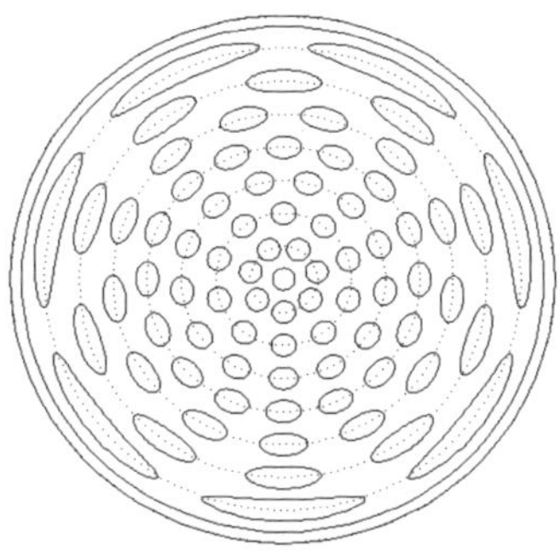

Figure 3.50. *Postel projection of the sphere*

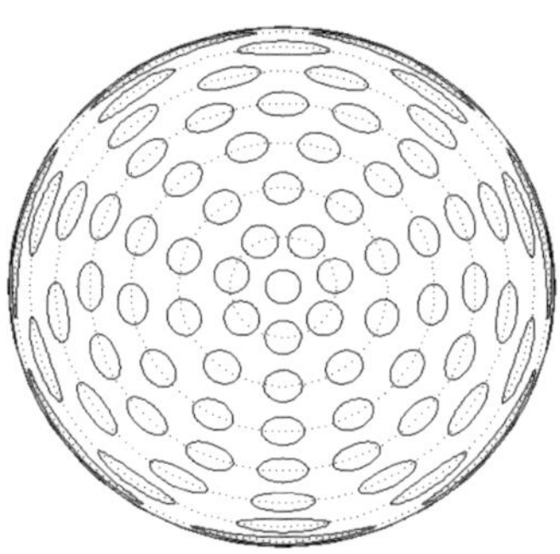

Figure 3.51. *Lambert equal-area projection of the sphere*

Rectangular maps in Cartesian coordinates and circular maps in polar coordinates	
Cartesian coordinates (x, y)	Polar coordinates (r, φ)
Plate carrée $\rightarrow$ $x = \varphi$, $y = \alpha$	Postel $\rightarrow$ $r = \theta$, $\varphi = \varphi$
Cylindrical Central $\rightarrow$ $x = \varphi$, $y = tg\alpha$	Cnomonique $\rightarrow$ $r = tg\theta$, $\varphi = \varphi$
Cylindrical Lambert $\rightarrow$ $x = \varphi$, $y = \sin\alpha$	Azimuthal Lambert $\rightarrow$ $r = 2\sin\frac{\theta}{2}$, $\varphi = \varphi$
Mercator $\rightarrow$ $x = \varphi$, $y = -\ln tg\frac{\theta}{2} = -\ln tg\left(\frac{\pi}{4}-\frac{\alpha}{2}\right)$	Stereography $\rightarrow$ $r = tg\left(\frac{\pi}{4}-\frac{\alpha}{2}\right) = tg\frac{\theta}{2}$, $\varphi = \varphi$
	Orthographic $\rightarrow$ $r = \sin\theta$, $\varphi = \varphi$

Table 3.5. *Comparison of rectangular and polar maps*

3.4.8. *Direct computation of azimuthal projections*

The representation of the full scene surrounding a point with the projection on the sphere needs a second projection to obtain a plane representation [BEC 11a]. In the case of the gnomonic projection, however, projecting on the sphere and then on the plane is the same as projecting directly on the plane because the center of the projection is the same.

For this projection, a straight line of the scene is transformed into a straight line in the plane projection. This characteristic is very useful because all of the polygons of the 3D space transform into polygons in the projection and the straight lines inside these polygons remain straight lines in their projection. Unfortunately, this projection is unable to completely represent the hemi-space surrounding the point.

Although it is not obtained in a single pass, the orthographic projection does not need extra computation after the projection on the sphere. The final result is obtained by removing the third component of the point projected on the sphere. The straight lines of the 3D space are transformed into great circle arcs on the sphere and, after the orthographic operation, into conical segments, more precisely ellipse arcs whose centers coincide with the center of the disk representing the projection of the full sphere. The interpolations necessary to draw the elements become rather complex.

The position of the point is described either in Cartesian or in cylindrical coordinates.

$$P = \begin{bmatrix} x_0 & y_0 & z_0 \end{bmatrix} \quad \Leftrightarrow \quad P = \begin{bmatrix} r_0 & z_0 & \varphi_0 \end{bmatrix} \qquad [3.46]$$

In the latter formulation, due to the radial symmetry, the coordinate φ_0 is not modified by the projection. The projection on the sphere is easily computed [3.1].

As proposed in [GAS 08], instead of projecting on the sphere, an azimuthal projection may be replaced by a projection onto a general nonlinear surface whose equation depends on the given projection. The azimuthal representations defined before with two successive projections are now computed throughout a single nonlinear one.

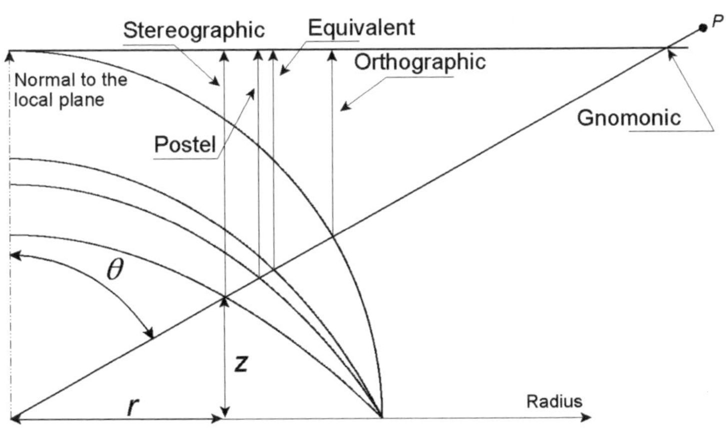

Figure 3.52. *Nonlinear projections for the different azimuthal projections*

Assuming that the point (r_S, z_S, θ_S) is defining the expected surface, the zenithal angle θ of the vector P, on the one hand, and the radial r and vertical z coordinates of its intersection with the axisymmetrical unknown surface (Figure 3.52, [BEC 11a]), on the other hand, are related by the formula:

$$z = \frac{r}{tg\theta} \qquad [3.47]$$

The result is obtained by projecting directly on this axisymmetrical surface defined by its section with a plane containing the symmetry axis.

The elimination of the colatitude θ between equation [3.47] and the definition of the azimuthal projection given in Table 3.3 (or [3.37] for the stereography) provides the equation of this surface. In the case of the stereography, the solution is a parabola.

$$z = \frac{1-r^2}{2} \qquad [3.48]$$

Its rotation around the *z*-axis defines a circular paraboloid; this projection is indeed called *parabolic projection* in [GAS 08].

The equations of the four surfaces used to compute these projections are presented in the first line of Table 3.6. The four azimuthal projections $[r\ \varphi]$ of a point P defined in polar coordinates $[r_0\ z_0\ \varphi_0]$ are shown in the second line. The cheapest calculation concerns the gnomonic projection while the most expensive calculation corresponds to the equivalent one. Any position of the point P, but different from those reported in the last line of Table 3.6, is valid. To complete the information, the Postel surface, which is not in the table, is in the representation of Figure 3.52. As expected from Figure 3.40, it lies between the Lambert equal-area and the stereographic projections. It is possible to invent any curve, a section of the axisymmetrical surface, to produce deformation of the space environment. Possible shapes could be ellipses, hemi-cubes [COH 85] or polygonal shapes like a very flat rectangle; the latter is able to produce something very close to the gnomonic projection.

	Stereographic	Equivalent	Orthographic	Gnomonic
Surf.	$2z + r^2 - 1 = 0$	$z^2\left(r^2 - 2\right) + \left(r^2 - 1\right)^2 = 0$	$r^2 + z^2 = 1$	$z = 1$
Proj.	$\left[\dfrac{r_0}{\lvert P\rvert + z_0}\quad \varphi_0\right]$	$\left[\sqrt{1 - \dfrac{z_0}{\lvert P\rvert}}\quad \varphi_0\right]$	$\left[\dfrac{r_0}{\lvert P\rvert}\quad \varphi_0\right]$	$\left[\dfrac{r_0}{z_0}\quad \varphi_0\right]$
Lim.	$z_0 \neq -\lvert P\rvert$	$\lvert P\rvert \neq 0$	No	$z_0 \neq 0$

Table 3.6. *Four azimuthal projections [r φ] of the point P = [r0 z0 φ0]*

3.5. Pseudocylindrical projections

The projections presented below are widely discussed in [SNY 93], where they are referred to as *pseudo-cylindrical* in their use in mapping.

3.5.1. *Coordinates transformation from direct to transversal aspect*

By placing the reference point (or principal point) at the intersection of the equator and the prime meridian (Greenwich), we first establish formulas for azimuthal projections (Figure 3.53 in terms of longitude φ, latitude α and colatitude θ). The principal point initially located at the pole is moved to the intersection of the Greenwich meridian and the equator (geographical coordinates $\alpha = 0$, $\varphi = 0$).

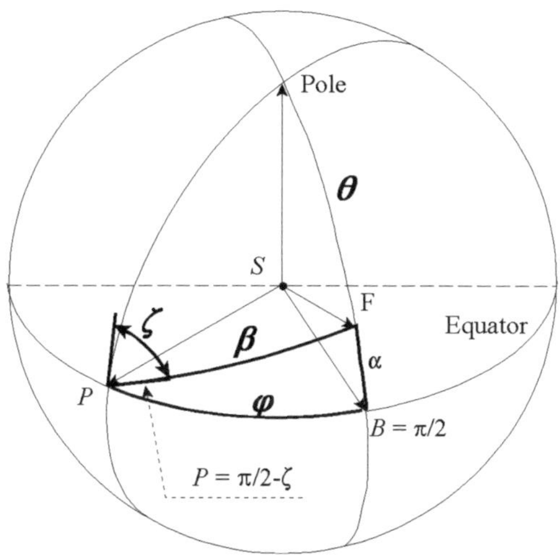

Figure 3.53. *Moving the principal point from the pole to a point P on the equator*

The reasoning is based on the formulas of spherical trigonometry. We finally obtain:

$$\begin{aligned} \cos\beta &= \cos\alpha\cos\varphi = \sin\theta\cos\varphi \\ tg\varsigma &= \text{cotg}\,\alpha\sin\varphi = \text{tg}\,\theta\sin\varphi \end{aligned} \qquad [3.49]$$

These two equations express how the principal point moves from the pole to the equator on the point of zero longitude. The angles β and ς are similar to the colatitude θ and longitude φ.

Inverse relations of [3.49] are:

$$\begin{aligned} \cos\alpha &= \sqrt{\cos^2\beta+\sin^2\beta\sin^2\varsigma} \\ \cos\varphi &= \frac{1}{\sqrt{1+tg^2\beta\sin^2\varsigma}} \end{aligned} \qquad [3.50]$$

Other fundamental formulas are:

$$\begin{aligned} \sin(arctg\ x) &= \frac{x}{\sqrt{1+x^2}} \\ \cos(arctg\ x) &= \frac{1}{\sqrt{1+x^2}} \\ \sin(\arccos x) &= \sqrt{1-x^2} \end{aligned} \qquad [3.51]$$

3.5.2. *Hammer projection*

Historically, the first development of this type is due to Aïtoff and concerns the transformation of the Postel projection. The method consists of dividing the longitudes by 2, then performing an *ad hoc* projection in its transverse (or equatorial) aspect and, finally, multiplying the abscissa by 2.

Step 1 : divide longitudes by 2 $\lambda = \varphi / 2$.

Step 2 : build one of the azimuthal projections, for instance, Postel **Error! Reference source not found.**:

$$\begin{aligned} X &= \beta\sin\varsigma \\ Y &= \beta\cos\varsigma \end{aligned}$$

Step 3 : multiply the x component by 2

$$\begin{aligned} x=2X &= 2\beta\sin\varsigma \\ y=Y &= \beta\cos\varsigma \end{aligned} \qquad [3.52]$$

By applying equations [3.49], where the longitude φ is replaced by the transformed longitude λ (Figure 3.53), we first obtain the following relations where both angles α and λ vary between $-\pi/2$ and $+\pi/2$.

$$Aïtoff \rightarrow \begin{array}{rcl} x & = & 2\arccos(\cos\alpha\cos\lambda)\sin(\text{arctg}(\text{cotg}\alpha\sin\lambda)) \\ y & = & \arccos(\cos\alpha\cos\lambda)\cos(\text{arctg}(\text{cotg}\alpha\sin\lambda)) \end{array} \quad [3.53]$$

By [3.51], we obtain a similar formula:

$$\begin{array}{rcl} x & = & 2\arccos(\cos\alpha\cos\lambda)\dfrac{\sin\lambda}{\sqrt{tg^2\alpha+\sin^2\lambda}} \\ y & = & \arccos(\cos\alpha\cos\lambda)\dfrac{tg\alpha}{\sqrt{tg^2\alpha+\sin^2\lambda}} \end{array} \quad [3.54]$$

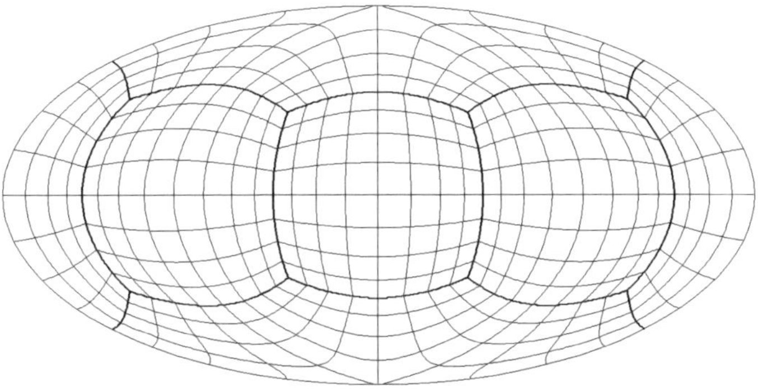

Figure 3.54. *Hammer projection of an 8 × 8 × 8 graticulated cube*

Instead of using the Postel projection, we take the equivalent Lambert azimuthal one [3.45], obtaining a similar elliptical projection called Hammer [LEI 55]. It is just necessary to replace the angle β of the formula [3.52] by sinus $\beta/2$ and scale the x component by a factor 2 and the y component by a factor π):

$$Hammer \rightarrow \begin{array}{rcl} x & = & \pi\sqrt{2}\sin\dfrac{\beta}{2}\sin\varsigma = \dfrac{2\pi\cos\alpha\sin\lambda}{\sqrt{1+\cos\alpha\cos\lambda}} \\ y & = & \dfrac{\pi}{\sqrt{2}}\sin\dfrac{\beta}{2}\cos\varsigma = \dfrac{\pi\sin\alpha}{\sqrt{1+\cos\alpha\cos\lambda}} \end{array} \quad [3.55]$$

The main advantage of this projection is that it provides a representation of the sphere, both complete and not too deformed. In Figure 3.54, a full hexahedral grid has been projected onto a sphere with the same center and then subjected to pseudo-cylindrical processing. The edges of the cube are in bold. In this projection, the rear face of the cube is divided into two parts, respectively, on the extreme left and the extreme right of the drawing. However, side square areas are not equal because they do not have the same solid angle, even if the cube squares that generate them are all equal.

In the Hammer representation, the projections of the six faces of the cube have the same area on the ellipse because they have the same solid angle on the sphere (the projection of the rear face of the cube is divided into two parts: in the extreme right and the extreme left).

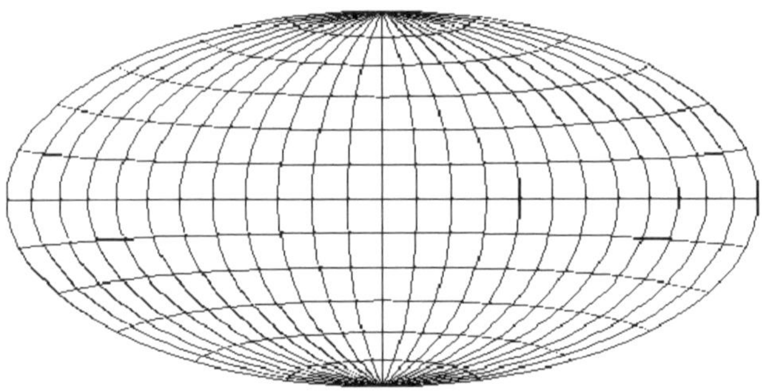

Figure 3.55. *Hammer, canvas, graticule 15°*

3.5.3. *Mollweide projection, another pseudo-cylindrical projection*

The same technique can be applied to the orthogonal projections: contraction of the sphere on a hemisphere, followed by a transverse azimuthal projection and then expansion of the projection on the axis corresponding to the previous angular contraction. But it is more interesting to perform a comparison with the Mollweide formulation [LAP 11].

We start from the formula:

$$\begin{aligned} x &= \frac{2\sqrt{2}}{\pi}\varphi\cos\beta \\ y &= \sqrt{2}\sin\beta \end{aligned}, \quad 2\beta+\sin 2\beta=\pi\sin\alpha \qquad [3.56]$$

The angle β is calculated iteratively. First, we calculate $\beta'=\alpha$, then:

$$\begin{aligned} \delta\beta' &= -\left(\beta'+\sin\beta'-\frac{\pi\sin\alpha}{1+\cos\beta'}\right) \\ \beta &= \beta'/2 \end{aligned} \qquad [3.57]$$

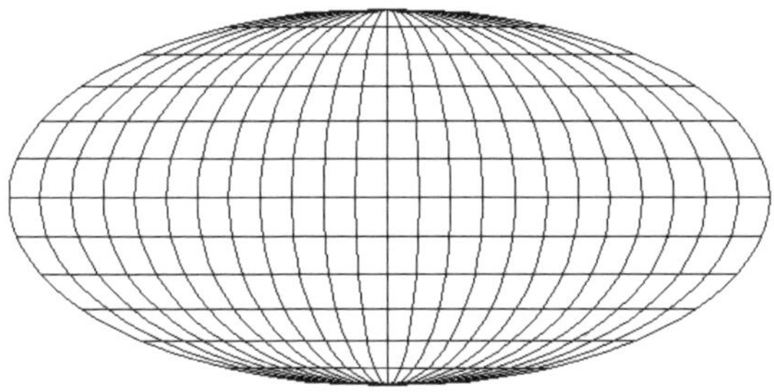

Figure 3.56. *Mollweide, canvas, graticule 15°*

3.6. Hemisphere tilling

3.6.1. *Presentation of the method*

In order to solve tessellation problems for the sphere or the hemisphere, it is affordable to start with the partition of a 2D image into equal-area cells. Both azimuthal and equal area on the cylinder projections (equivalent projections) are suitable to build equal-area partitions of the sphere or the hemisphere.

The projection on a cylinder is very easy to implement because it is rectangular. The meshing operations are obvious in the construction of a rectangular grid or a structured mesh [FRE 08], but the cells connected to the poles are transformed into triangles on the sphere (Figure 3.24–3.27).

In order to preserve an easy transformation between cells lying inside a disk and cells on the hemisphere, it is practical to keep the shape of their boundary curves as simple as possible. For this reason, these curves are restricted to straight lines and circular arcs. Therefore, the separation lines of the cells are concentric circles and radii in the azimuthal projections; they are meridians and small circles (parallel to the base) on the hemisphere.

To subdivide a disk, an obvious method consists of cutting it into sectors as in the usual *pie diagram*. However, when the number of pieces is substantial, the sectors become very thin. To limit their bad aspect ratio, it is possible to imagine a partition of each sector into equal-area cells. This method leads naturally to congruent meshes [VUE 06, ROS 04] (Figure 3.27).

If the congruency is not mandatory, another method can be implemented: the disk is divided into N circular rings [BEC 12a]. After the introduction of the ring I, whose external radius is r_i and internal radius is r_{i-1}, the number of cells inside the external circle is equal to k_i. The ring i is then defined by an equation expressing that two consecutive circles involve cells with the same area A_{cell}.

$$\begin{array}{lll} k_{i-1} \; A_{cell} & = & \pi r_{i-1}^2 \\ k_i \; A_{cell} & = & \pi r_i^2 \end{array} \quad \text{then: } \left(\frac{r_i}{r_{i-1}}\right)^2 = \frac{k_i}{k_{i-1}} \qquad [3.58]$$

The internal or the external radius of the ring is computed, given the other radius and the numbers of cells in the internal and external disk. The central region is a disk that includes k_0 (integer greater than zero) cells and whose radius r_0 is greater than zero. When going outward, by adding a ring, it is necessary to respect the condition $k_i > k_{i-1}$. When going inward, by removing a ring, the conditions are: $k_i \leq k_{i-1}$ and $k_i > 0$.

This rule allows adding at each step any number of cells and, in the opposite process, to remove no more than k_{i-1} cells. With [3.58], the cell areas are perfectly controlled, but their shapes are implicitly driven by the number of added (or removed) cells in the ring i. The problem is then: how to control their shapes?

The simplest way to characterize the shape of a quadrangular cell limited by two concentric circles and two radii is to introduce the idea of *coverage index* [BBE 13]. A measure of the aspect ratio of a cell defined on the sphere is the size of its greatest possible inscribed circular cap. In the finite element method, tests of inscribed circles in triangles also provide criteria to control the mesh uniformity [FRE 08]. If the smallest on the hemisphere of the greatest inscribed circular cap in each band is used to cover the full hemisphere, it is guaranteed that the caps are not overlapping. This procedure also provides the best coverage of the hemisphere within the existing cells. The coverage index I_c is then the ratio between the summation of the cap areas and the hemisphere area.

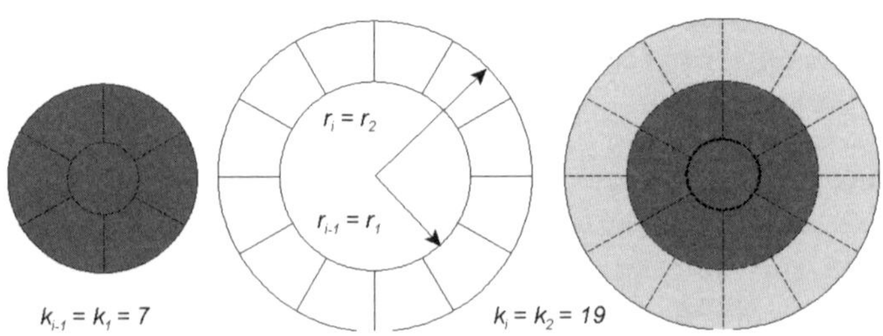

Figure 3.57. *Construction of a new ring*

As we know that the aspect ratio of a rectangle is the height to width ratio, we can say in a first approximation that, for a cell inside a disk, it is the ratio between the mean circumferential dimension (mean radius multiplied by the angular aperture) and its radial dimension (difference between the external and internal radii).

$$a_i = \frac{2\pi}{k_i - k_{i-1}} \frac{r_i + r_{i-1}}{2} \frac{1}{r_i - r_{i-1}} \qquad [3.59]$$

Due to the constant area constraint, all the aspects ratios of the cells pertaining to the same ring are identical. By [3.58], it is possible to eliminate the radii and to express the aspect ratio as a function of the numbers of cells inside two consecutive disks.

$$a_i = \frac{\pi}{\left(\sqrt{k_i} - \sqrt{k_{i-1}}\right)^2} \qquad [3.60]$$

This formula provides the partitioning constraint for obtaining cells with an imposed aspect ratio in the ring i. The aspect parameter $p_i = \pi / a_i$ is now used to express the condition on aspect ratio.

$$\sqrt{k_i} = \sqrt{k_{i-1}} \pm \sqrt{p_i} \quad \text{with: } p_i = \pi / a_i \qquad [3.61]$$

It gives the number of cells in a new disk according to the aspect parameter and the number of elements in the internal or external one. However, it is necessary to remember that the variables k_i are integers. Consequently, in general, the solutions of equation [3.61] violate slightly the aspect parameter constraint. Using formula [3.58] and an integer as close as possible to the solution of [3.61] makes it possible to compute the new radius. This means that the equal-area condition is fully satisfied while the aspect ratio constraint is approximated.

3.6.2. *Exact fulfillment of the aspect ratio constraint*

A careful analysis of equation [3.61] allows us to identify sequences of solutions leading to integer numbers. We call them "exact" solutions. A solution is exact if k_i and k_{i-1} (integers) satisfy [3.61], with an imposed aspect parameter p_i. There are several possibilities.

A first solution is obtained with a constant parameter p ($p_i = p$) and a sequence of square numbers. If p is also a square number ($p = m^2$), an exact solution is obtained.

If $p = m = 1$, the number of cells inside the circle i in a total of N is:

$$k_i = (i+1)^2 \ ; \ i = 0, N \qquad [3.62]$$

If $p = 4$, the sequence corresponds to the squares of the odd numbers. The aspect ratio is equal to $\pi/4 = 0.785$. This solution exhibits high symmetry. For $p = 9$, the successive circles contain $(1 + 3i)^2$ cells and the aspect ratio is equal to 0.349.

p	$a=\pi/p$	k_0	k_1	k_2	k_3	k_4	k_5	k_6	k_7	k_8	k_9	
1	3.142	1	4	9	16	25	36	49	64	81	100	[BEC 12a]
4	0.785	1	9	25	49	81	121	169	225	289	361	[GRI 92]

Table 3.7. *Cell distribution for different aspect parameters*

3.6.3. *Approximate fulfillment of the aspect ratio constraint*

The other way to solve the problem is to choose for k_i the closest integer of the solution of equation [3.61].

$$k_i = \langle\left(\sqrt{k_{i-1}} \pm \sqrt{p_i}\right)^2\rangle_{\text{integer}} \qquad [3.63]$$

Then, with a parameter $p = \pi$ (square cells), it is easy to create a tessellation with an arbitrary number of cells, for instance exactly 800 cells. In this case, the solution consists of 17 rings and the central circle. The accumulated sequence of cells in the rings is: [1 9 23 43 69 101 138 180 227 279 335 394 457 522 590 659 729 800].

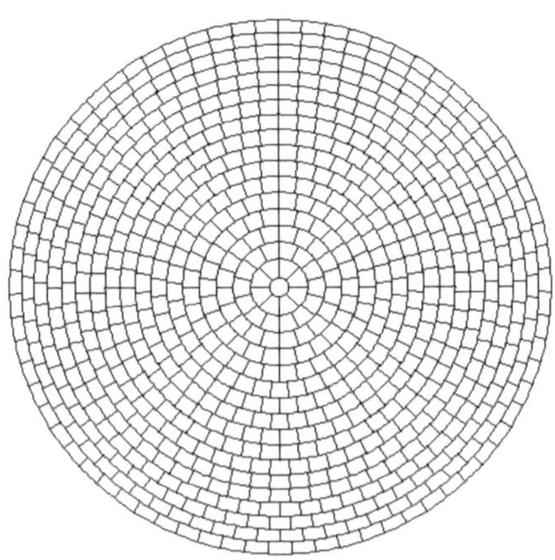

Figure 3.58. *800 equal-area cells, $p = \pi$, or $a = 1$; $k_{17} = 800$*

3.6.4. *Equal-area cells and constant aspect ratio on the hemisphere*

A point (θ, φ) on the hemisphere is projected in the position (r, φ) in the disk and, as computed before [3.45], the aspect ratio of a disk cell is:

$$a_{i\_disk} = \frac{S_{disk}}{(r_{i-1} - r_i)^2} \quad [3.64]$$

The problem of definition of equal-area cells on the sphere was recently analyzed and exploited [BEC 12a, LEO 06, FAU 08]. The aspect ratio of a spherical patch limited by two meridians and two parallels is obtained by dividing its area by the square of the length of the arcs of meridians. The indices r and l or t and b are used to define the boundaries: right, left, top and bottom:

$$a_{i\_sphere} = \frac{S_{sphere}}{(\vartheta_b - \vartheta_t)^2} = \frac{(\varphi_r - \varphi_l)(\cos\vartheta_b - \cos\vartheta_t)}{(\vartheta_b - \vartheta_t)^2} \quad [3.65]$$

If the colatitude θ is close to 90°, this equation gives the aspect ratio of a small rectangular patch close to the base of the hemisphere.

The situation on the unit hemisphere is related to its azimuthal equal-area Lambert projection by [3.44].

Taking into account the equality of the areas of a cell and of its projection, the ratio between the two aspect ratios becomes:

$$\frac{a_{i\_sphere}}{a_{i\_disk}} = \frac{S_{sphere}}{(\vartheta_{i-1} - \vartheta_i)^2} \frac{(r_{i-1} - r_i)^2}{S_{disk}} = \left(\frac{r_{i-1} - r_i}{\vartheta_{i-1} - \vartheta_i} \right)^2 \quad [3.66]$$

The test presented in Table 3.8 corresponds to a dome containing twice the number of cells presented in [TRE 87] where it was composed of 145 cells, 144 in the rings and 1 in the top spherical cap. We then have 288 cells and one in the top spherical cap.

The table contains, in the first line, the numbering of the rings on the disk or the bands on the sphere from top to bottom; in the second line, the cumulated number of cells; in the third line, the aspect ratio of the disk cells of the Lambert equal-area projection; in the fourth line, the aspect ratio on the sphere; and in the fifth line, the coverage index in each band.

The coverage index of a cell is the ratio between the area of its greatest inscribed circle and the area of the cell [BEC 12a].

1	i	1	2	3	4	5	6	7	8	9	10	11
2	k_i	1	8	21	40	64	93	127	164	204	246	289
3	a_{i_disk}	1.01	1.07	1.10	1.15	1.20	1.27	1.38	1.50	1.66	1.87	2.15
4	a_{i_sphere}	1	1.05	1.05	1.05	1.05	1.04	1.05	1.05	1.05	1.04	1.04
5	*areas*	0.79	0.73	0.77	0.77	0.77	0.78	0.76	0.78	0.77	0.77	0.77

Table 3.8. *Description of the 289 cells dome*

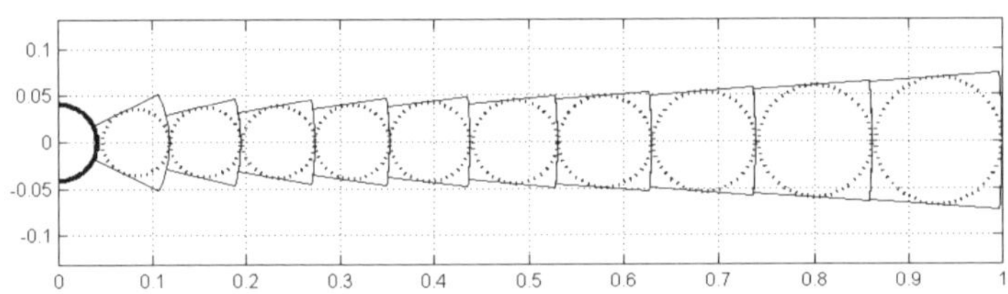

Figure 3.59. *Stereographic projections of cap bases drawn in a sector, 289 cells, Ic =0 .73*

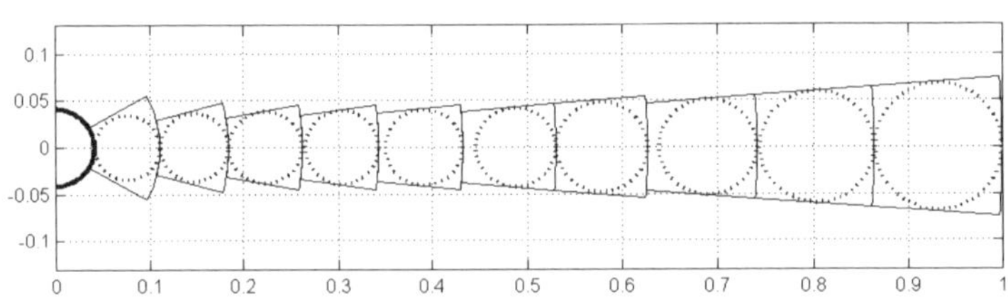

Figure 3.60. *Stereographic projections of cap bases (high symmetry), 289 cells, Ic = 0.68*

If we force the cell sequence to contain a multiple of 6 in each ring, we lose quality, mainly in rings 5, 6 and 8, but we obtain the required symmetry. We observe that the solution exhibits a better coverage index than in the 145 cells dome (Figure 3.62).

Using equation [3.60] that defines the aspect ratio on the circle, the expression becomes:

$$a_{i\_sphere} = \pi \left(\frac{r_i - r_{i-1}}{\sqrt{k_i} - \sqrt{k_{i-1}}} \frac{1}{(\vartheta_i - \vartheta_{i-1})} \right)^2 \qquad [3.67]$$

Using relations [3.58] and [3.45]:

$$\vartheta_i = \vartheta_{i-1} - \frac{2}{a_{i\_sphere}} \sin \frac{\vartheta_{i-1}}{2} \sqrt{\frac{\pi}{k_{i-1}}} \qquad [3.68]$$

After computing the zenithal angle of a new ring, the new radius is deduced from [3.45] and the number of cells in the internal circle is deduced from [3.58]. For a test with 289 cells distributed with different levels of symmetry, the results are given in Figures 3.61 and 3.62.

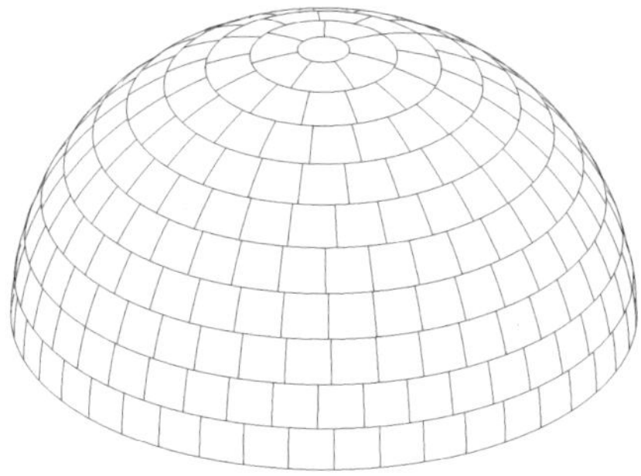

Figure 3.61. *Dome 289 cells, Ic = 0.73*

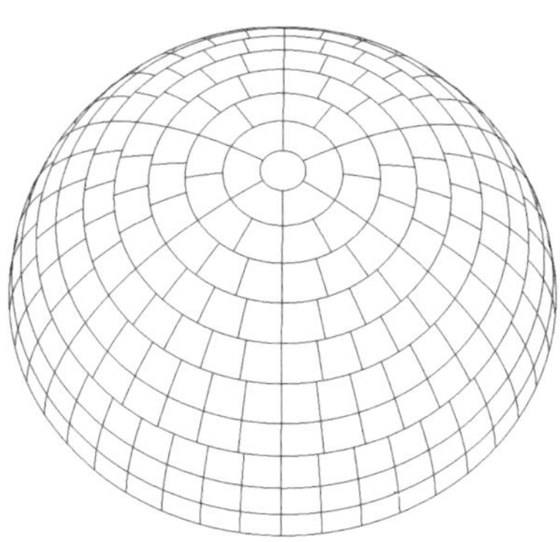

Figure 3.62. *High symmetry, 289 cells, Ic = 0.68*

3.6.5. *Conclusion*

The partitions of the circle or the hemisphere are performed with quadrangular cells enclosed in the disk by concentric circular arcs and radii and on the hemisphere by meridians and small circles. In this framework, it is straightforward to map the hemisphere on an azimuthal projection or inversely. Defining all the cells with equal areas in a circle, it is simple to add or remove a ring imposing either the number of cells or their aspect ratio. The same holds for the hemisphere.

According to the proposed method, all the schemes proposed in the literature are easily reproduced [BJO 03, CRI 00] but many new original solutions are also encountered. This method is providing attractive flexibility and generality for introducing tessellations on spheres, hemispheres or circles.

A further analysis of the solutions also shows that the transformations between the hemisphere and the circle or *vice versa* do not maintain the aspect ratios of the cells. However, this property was introduced without difficulty as a constraint in the process of construction of the circular or spherical set of equal-area cells [BEC 12a].

4

Geometry and Physics: Radiative Exchanges

4.1. Geometric wave propagation

The human body perceives mainly three types of radiation: sound, light and heat. The fact that these energies are radiated in space implies a strong relationship with the geometry of the scene where they propagate. This is why geometrical theories of physical radiation have appeared quite naturally: geometrical optics (visual rays, from the time of Euclid, and light rays), geometrical acoustics (sound rays) and, finally, and less naturally, geometric modeling of thermal radiation (radiosity equation).

Planck's black body law shows that the Sun, which has a surface temperature of about (5430ºC), emits virtually all its energy below 4 μm (shortwave), while the Earth, with an average temperature of 288 K (15°C), almost exclusively radiates above 4 μm (longwave). Shortwaves arriving to us from the Sun and sky are practically decoupled from longwaves emitted by the objects that surround us. In other words, the objects of the scene capable of emitting shortwave are exceptional: this is basically the lamps, and objects brought to high temperatures incandescence). Under normal conditions, shortwaves do not depend on the surface temperature of the scene.

With a speed of about 300,000 km/s, light propagates almost instantaneously at our scale. In general, air can be considered as perfectly transparent. This is no longer the case for very long distances or in the presence of clouds, fog or smoke. This is called the "participatory environment". On clear days, between 80 and 90% of the sunlight reaches us directly (solar beam). The rest is mainly distributed through the sky, which behaves as a secondary source (diffuse component). If a cloud passes over the sun, the direct component disappears, and the diffuse component increases significantly.

When they meet the surface of a black body at room temperature, short waves are completely absorbed and re-emitted as long waves. Re-emission does not interfere in the shortwave balance. On the objects in the scene that are not black bodies, a portion of the light is reflected. If these objects are not opaque, another part of the light is refracted.

Each surface has a law of reflection, which depends not only on material properties but also on the incident wavelength. The two extremes are the specular reflection and the Lambertian reflection (perfectly diffuse). The first needs surfaces that are smooth on the scale of the wavelength in question. For light, with a few hundred nanometers, perfect mirrors are rare and, as a first approximation, we can often assume that the scattering is Lambertian on all objects in the scene. For acoustics, where the wavelengths range from a few centimeters to a few meters, the specular reflection is the most common; so it is very difficult to get perfect diffusers for all frequencies.

When we look at the shortwaves as a whole (e.g. for the evaluation of solar thermal contributions on the building), a single reflection factor, averaged over all frequencies, is generally considered for each material. Sunlight contains ultraviolet (UV rays, below 380 nm), visible light (between about 380 and 760 nm) and near-infrared (between 760 nm and 4 μm). The human eye is sensitive only to the second part of the spectrum, according to a bell curve that peaks at 555 nm. Filtering shortwave by the sensitivity curve of the human eye, one shifts from radiometric to photometric magnitudes. Other applications require additional filters: the sensitivity curve for the photosynthesis of plants (the same frequency range as for the eye, but with a maximum at the two extremes and a minimum at the center corresponding to the green light, which is reflected) or solar cells (with a wider range of sensitivity, including a portion of the near-infrared).

Synthesis images in color can be realized by simulating only three frequency bands (RGB), which correspond to three types of cone, the photosensitive cells lining the retina of the human eye. This system does not reproduce all the colors [FOL 90], but gives these images of virtual scenes the same quality as digital pictures.

The simplest technique involves two steps. First, the visual rays are sent deterministically from the eye to each pixel of the screen, and these rays will then capture a perspective view of the scene. For each element of the scene appearing in this perspective, a test is performed to check whether it sees each of the lamps. To each lamp viewed, a light contribution will be added, which is a function of the square of the distance between the element and the lamp, the incidence angle of the light beam on the element, the intensity emitted by the lamp in this direction and the color of the item.

Since the 1990s, realistic renderings could thus be made on PC. For a quick calculation, the illumination of surfaces was merely simulated without taking into account the reflection. A mirror could still be added at a reasonable extra cost because the specular reflection does not imply more than a single reflected ray for a given incident beam. In contrast, the diffuse reflection posed a serious problem because, for an incident ray, it is necessary to cast a large number of reflected rays in all directions. In lighting calculation software, this technique has been established, but it must be limited to the calculation of the first or second reflection. For each added order of reflection, computation time increases exponentially.

In the early 1990s, new algorithms, based on *radiosity*, have increased. This method, which immediately gives the infinite reflections, is particularly well suited for an instantaneous phenomenon such as light. Moreover, the diffuse reflection is, this time, the easiest to calculate. In section 4.3, we will detail the radiosity equation.

At the level of the architecture, the sound field diverges from the light field on two very important points. First, as mentioned above, the reflection is primarily specular. On the other hand, the speed of sound in air is about 340 m/s, and it cannot at all be considered infinite. Compared to the direct sound, reflections arrive at the ear with delays that are functions of differences in path. When the difference reaches 34 m (i.e. a delay of 100 ms), an isolated reflection can be perceived as an echo. In a closed room, the reflections are numerous, and they form the phenomenon of reverberation.

The basic method for architectural acoustics is random ray tracing, according to the Monte Carlo method. In section 4.4, we will describe some useful procedures for acoustic and light simulation, based on the previous chapter's results.

The thermal radiation (longwave) is more complicated because it depends on the surface temperatures of the scene, and it is strongly coupled with two other forms of transmission: thermal conduction and convection. According to the Stefan–Boltzmann law, each body radiates an amount of energy proportional to the fourth power of its surface temperature.

Today, thermal cameras allow us to visualize the strange world of infrared. In these images, past mixes with present. Traditional photography, shortwave, had accustomed us to the instantaneous snapshot. Delays and focalizations of linear acoustics are purely geometric. However, they are enough to surprise us (whispering echo, fluttering echo, etc.). In the presence of wind and even more of a temperature gradient, nonlinear acoustics can surprise us more by bringing us suddenly, over obstacles, very distant sounds. But the infrared is something else. You can see traces of your footsteps, heat of the previous day emerging from a stone wall, warm clouds scroll through an icy sky, etc.

A blue summer sky, so bright in the visible, emits virtually nothing in longwaves, and the equivalent temperature of the sky may fall to –50°C at its zenith, while it remains up to 0° near the horizon (the atmosphere layer is thicker there) or under an overcast sky. We can considerably cool a bottle of water by exposing it to the summer sky, as long as it and the surfaces that surround it are protected from the Sun.

Visual factors are of great importance in exchanges by thermal radiation, and it is precisely for thermal applications that the radiosity method was first developed in the 1950s, well before it was taken successfully for simulating light.

While light, acoustic and thermal radiations are very different in their physical nature, their production and their perception by our sense organs, it is nevertheless true that they are united by their geometric behavior and that the same methods (ray tracing, radiosity) enable their simulation.

For some scientific communities, such as computer graphics, numerical methods, and environmental, urban and architectural physics, the integration of the different forms of perceived radiation has become a necessity.

However, the unification of formulations remains to be done. To give just one example, acoustic intensity is expressed traditionally in Wm^{-2}, while light intensity is in Wsr^{-1}.

In Chapter 1, we recalled what is shared by our perceptions of music and light (relativity, within-time structure). In Chapter 2, we have emphasized the main property of the foreshortening. In Chapter 3, we have developed widespread geometric concepts (the solid angle, the form factor). In this chapter, we now try to give some ideas to achieve a comprehensive understanding of the physical phenomena related to radiation.

4.2. The radiosity equation

In this section, the physical quantities used to calculate the light and thermal radiation exchanges are introduced. We continue to rely on purely geometric concepts introduced in previous chapters, but combining them with physical data, which in the context of this book are linked to the characteristics of surfaces. With these data, we can implement the methods for calculating radiative exchanges [COH 85, KAJ 86, GRE 86, HEC 92, HEC 93, COH 93].

The essential characteristics of the surfaces are related to their emission, reflection and absorption properties.

First, let us recall the definition of a solid angle. At a given point *X*, the solid angle of an object is the area of its projection on the unit sphere surrounding the point. If the object is a differential area, i.e. dA_Y, its projection on the sphere, called a differential solid angle, is given by:

$$d\omega_{x-dA_y} = \frac{\cos\vartheta_y \, dA_y}{r^2} \qquad [4.1]$$

In this expression, θ_Y is the angle between the normal to the differential area and the line connecting it to the point located at distance *r*.

4.2.1. *Surface sources*

Let us consider a differential element *dA* pertaining to the radiating surface of an object. The total energy flux *dP*, leaving the element surface *dA* into the surrounding hemisphere, is called *radiant power* or *energy flux* (watts) [VOL 10].

When relating this radiant power to the emitting surface area, we define the *exitance H*, in W m^{-2} (also called *emittance* or *emissive power*; the *Commission Internationale de l'Eclairage* recommends the name *exitance* instead of *energetic emittance*):

$$H = \frac{dP}{dA} \qquad [4.2]$$

Conversely, if we consider the total incident power from a hemisphere on a surface *dA*, the same definition holds for the *irradiance E*:

$$E = \frac{dP}{dA} \qquad [4.3]$$

Exitance and irradiance refer to the same physical quantity and, therefore, are expressed in the same unit (W m^{-2}), but the corresponding energy flux is either emitted or received by *dA*.

The *radiant intensity* I is the radiant power that is emitted from a point source of a radiating object into a solid angle element $d\omega$ in a given direction, characterized by the angles (θ, φ). It is expressed in watt/steradian (W sr^{-1}).

$$I = \frac{dP}{d\omega} \qquad [4.4]$$

Radiant intensity is related to the most usual quantity of radiometry: the *radiance* L, which is used to characterize extended sources [COH 93]. It is defined as the amount of radiant power per unit of projected source surface and per unit of solid angle.

$$L = \frac{d^2 P}{\cos\theta \, d\omega \, dA} \quad \rightarrow \quad d^2 P = L \cos\theta \, d\omega \, dA \qquad [4.5]$$

The significance of this definition becomes clearer when the equation is used to compute the total radiant power:

$$P = \iint L \cos\theta \, d\omega \, dA \qquad [4.6]$$

The total radiant power results from adding up radiance contributions over the area and the solid angle of the hemisphere. If only the integration over the solid angle is performed, we obtain the exitance [4.2].

$$H = \frac{dP}{dA} = \int L \cos\theta \, d\omega \qquad [4.7]$$

Alternatively, integration over surface area results in radiant intensity [4.4].

$$I = \frac{dP}{d\omega} = \int L \cos\theta \, dA \qquad [4.8]$$

The geometrical factor cos θ can be easily understood. Any emitting surface area dA is observed to be the largest in the direction perpendicular to the surface. For any other direction, only the projection of dA perpendicular to this direction contributes to the emitted radiation.

Hence, radiance is a measure for the radiant power of an emitter with surface area dA that passes through a surface normal to the emission direction. Since this surface defines a solid angle in this direction [4.1], radiance is a true measure of the amount of radiation that is emitted into a certain direction and solid angle.

The difference between radiant intensity and radiance becomes more evident for the so-called *Lambertian radiators* [VOL 10]. A Lambertian radiator is a theoretical device that emits or reflects radiance independent of the angle, i.e. it emits uniformly into the hemisphere. Good examples found in the real world are blackbody sources or perfect diffusely scattering surfaces. In these cases, L is constant and:

$$I = L\,A\,\cos\theta = I_0 \cos\theta \qquad [4.9]$$

The radiance is constant (its product by the area A is denoted by I_0) as in the left part of Figure 4.1, while the intensity depends on the orientation as in the right part of the same figure.

Radiance is directional like a ray, but unlike the ray, it is not a discrete variable; we should rather interpret it as a pencil. The radiant power is proportional to the radiance, to the surface projected on the direction and to the opening of the pencil. If the radiance of the pencil and the opening are constant, the transmission takes place according to the diagram (Figure 4.1(b)) with a maximum in the direction normal to the surface element and a zero value for a pencil tangential to the surface element.

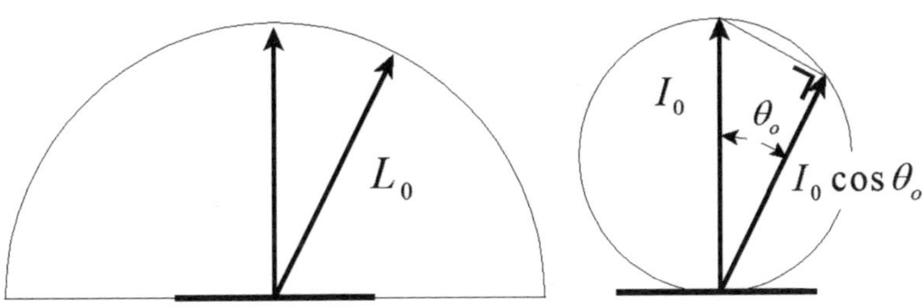

Figure 4.1. *Radiance (left); radiant intensity as a function of the direction of emission (right)*

Radiosity characterizes the total radiation leaving the surface; it is composed of emittance and reflected radiation. The corresponding quantity characterizing the radiation that reaches the surface is the irradiance [4.3].

The energetic equilibrium of a set of radiant surfaces is expressed by the following integral equation called the *global illumination equation*:

$$L_o(\theta_o,\phi_o,x) = L_e(\theta_o,\phi_o,x) + \int_{\Omega_i} \rho_{bd}\; L_i(\theta_i,\phi_i,x)\,\cos\theta_i\, d\omega_i \qquad [4.10]$$

Equation [4.10] models the energy balance at point x, since it expresses the outgoing radiance L_o according to different radiant energy sources (the indices i and o correspond to *in* and *out* radiances). The first term on the right corresponds to the surface own emission. It is different from zero only for the surfaces that emit radiation. For light, this is only the case for the light sources. For heat transfer simulations, all surfaces produce radiation depending on their temperature and emissivity.

The second term on the right represents the reflection: the reflected radiance is the integral over all directions of the product of the differential incident irradiance $L_i \cos\theta_i \, d\omega$ by a *bidirectional reflectance distribution function* (BRDF).

Equation [4.10] is an integral equation: the unknown radiance appears in both the left side and under the integral sign on the right side. This type of equation can rarely be solved analytically.

4.2.2. *Lambert diffuse reflection*

Let us consider the case of Lambert diffuse reflection, where the radiation is scattered equally in all directions, regardless of the incident direction. In other words, in this *perfect diffuse* case, the BRDF is equal to a constant and it depends only on the position.

Assuming that the surface does not emit and the reflectance depends only on the position, we can write:

$$\begin{aligned} L_{o,L} &= \int_{\Omega_i} \rho_L(x) L_i(\theta_i, \phi_i, x) \cos\theta_i d\omega \\ &= \rho_L(x) \int_{\Omega_i} L_i(\theta_i, \phi_i, x) \cos\theta_i d\omega \\ &= \rho_L(x) E(x) \end{aligned} \quad [4.11]$$

The quantity *E(x)* [4.3] represents the total irradiance reaching the surface at point *X*.

This brings three conclusions:

1) The reflected radiance is proportional to the incident irradiance.

2) The reflected radiance is constant and thus the same in all directions, since neither E nor ρ_L depends on (θ_0, φ_0). It is completely independent of the incident radiation.

$$L_{o,Lambert}(\theta_o, \phi_o, x) = L_{o,Lambert}(x) \quad [4.12]$$

3) Taking into account the inclination of the radiating direction, the transmitted power corresponds to the drawing of Figure 4.1.

By integrating the outgoing radiance, we obtain the definition of the *reflectance*:

$$\begin{aligned}\rho &= \frac{\int_{\Omega_0} L_{o,L}\cos\theta_o d\omega}{\int_{\Omega_i} L_i \cos\theta_i d\omega} \\ &= \frac{L_{o,L}(x)\int_{\Omega_0}\cos\theta_o d\omega}{E} \\ &= \frac{\pi L_{o,L}(x)}{E} \\ &= \pi\rho_L\end{aligned} \qquad [4.13]$$

When the BRDF is constant, the reflectance is also constant. Moreover, this relationship can be used to write the BRDF as a function of the reflectance: $\rho_L = \rho / \pi$. Here, since the outgoing radiance is constant, the radiosity corresponding to the reflection is given by:

$$B(x) = \int_{\Omega} L_{o,L}\cos\theta_o d\omega = \pi L_{o,L}(x) \qquad [4.14]$$

In the case of ideal diffuse surfaces, radiance and radiosity are therefore interchangeable.

Combining [4.11] and [4.13], we obtain:

$$\rho = \frac{B}{E} \qquad [4.15]$$

For diffuse reflection, the reflectance is equal to the radiosity divided by the irradiance.

Equation [4.10] is then:

$$L_o = L_e + \frac{\rho(x)}{\pi}\int_{\Omega_i} L_i \cos\theta_i \, d\omega \qquad [4.16]$$

From the definition of the incident irradiance:

$$E(x) = \int_{\Omega_i} L_i \, \cos\theta_i \, d\omega \qquad [4.17]$$

The reflected radiance E does not depend on the direction. If the same is true for the emitted radiance L_e, we have:

$$L_o(x) \;=\; L_e(x) \;+\; \frac{\rho(x)}{\pi}\,E(x) \qquad [4.18]$$

and multiplying by π:

$$B(x) \;=\; H(x) \;+\; \rho(x)\,E(x) \qquad [4.19]$$

The first term of the second member $H\ (x)$ represents the emittance, i.e. the radiation produced by the surface.

4.2.3. *Interactions between surfaces*

Equation [4.19] is difficult to solve because E depends on the radiosities of the other surfaces. Let us transform expression [4.17] into a surface integral. The invariance of radiance [KAJ 86] along the line of sight x-y (Figure 4.2) gives:

$$L(\theta_x,\varphi_x,x) \;=\; L(\theta_y,\varphi_y,y) \;=\; \frac{B(y)}{\pi} \qquad [4.20]$$

Developing the differential solid angle $d\omega_Y = \dfrac{\cos\theta_Y dA_Y}{r^2}$, and integrating not on the solid angle but on the surface (the symbol dA_Y and the limits of integration A_Y are used, the index indicating which surface is concerned), equation [4.17] becomes:

$$E(x) \;=\; \frac{1}{\pi}\int_{A_y} B(y)\,\frac{\cos\theta_x\ \cos\theta_y}{r^2}\,V(x,y)\,dA_y \qquad [4.21]$$

The function $V(x,\ y)$ is introduced to characterize the visibility; it is equal to 1 if point X sees point Y; otherwise, it is zero.

The radiosity equation becomes:

$$B(x) = H(x) + \frac{\rho(x)}{\pi}\int_{A_y} B(y)\,\frac{\cos\theta_X\ \cos\theta_Y}{r^2}\,V(x,y)\,dA_y \qquad [4.22]$$

This relationship is valid for exchanges between two surfaces A_x and A_y. It can be solved for simple geometric configurations. Inside the integral [4.21], there is a quantity called *view factor* (or *form factor*). It remains to show how to remove the term B from the integral. This operation is performed by discretizing the equation.

4.2.4. *Discretization of the radiosity equation*

Following the terminology used in the field of CAD [FOL 90, FAR 90], a *patch* is a geometric item limited mostly by four sides and defined in parametric form. This surface is of any shape; it is not necessarily planar and is *delimited* by non-planar curves. The extension to NURBS of the polynomial functions [VER 75] allowed us to unify the mathematical formulation of the conical curves through the homogeneous coordinates.

Discretizing the scene leads us to the following assumptions. The CAD patches are divided into elements (or tiles) in which we assume that the radiosity is constant. Radiosity is approached on each tile by a piecewise constant function, like in Galerkin or finite element formulation [FRA 72, ZIE 71].

In mathematics, in the field of numerical analysis, Galerkin methods allow the transformation of a continuous operator problem (e.g. PDEs) into a discrete problem. Certain constraints on the space of functions are applied to define a space with a finite number of base functions. The finite element method is one of the Galerkin methods.

The reflectance is assumed to be constant on all elements. The scene is described by a set of N elements A_j (j=1, N) for which the differential surface is denoted as dA_j. The variables related to the elements are identified by their index.

Equation [4.22], which sets the relationship between the surface element containing X and the other element containing Y, is transformed into a discrete system. This step is very important because it allows us to solve the system very easily. The main characteristic is that the mesh to be created need not necessarily be congruent: it must cover the entire area, but the common borders are free of any constraint. This facilitates the mesh generation. In addition, it opens the door to opportunities for local refinement of the mesh and thus hierarchical methods [AUP 93, BES 05].

First, we write the relation between the patch containing X and the N others containing the Yj points. For this, we replace the right member of [4.22] by a sum of integrals:

$$B(x) = H(x) + \rho(x) \sum_{j=1}^{N} \int_{A_j} B(y_j) \frac{\cos\theta_i \cos\theta_j}{\pi r^2} V(x, y_j)\, dA_j \qquad [4.23]$$

Because each element has a constant radiosity, it can leave the integral symbol:

$$B(x) = H(x) + \rho(x)\sum_{j=1}^{N} B_j \int_{A_j} \frac{\cos\theta_i \cos\theta_j}{\pi r^2} V(x, y_j)\, dA_j \qquad [4.24]$$

This equation suggests a fine point. Indeed, the left side variable *B (X)* varies when the points move on the involved elements. This is paradoxical if we make the assumption of the constancy of the radiosity on each element. We can eliminate this paradox by assuming that it is constant only in the process of radiation to other surfaces. This becomes:

$$B_i = \frac{1}{A_i} \int_{A_x} B(x)\, dA_x \text{ avec } A_i \equiv A_x \qquad [4.25]$$

and:

$$H_i = \frac{1}{A_i} \int_{A_x} H(x) dA_x \qquad [4.26]$$

Finally:

$$B_i = H_i + \rho_i \sum_{j=1}^{N} B_j \frac{1}{A_i} \int_{A_j} \int_{A_i} \frac{\cos\theta_i \cos\theta_j}{\pi r^2} V(x_i, x_j)\, dA_i dA_j \qquad [4.27]$$

More concisely, we write the radiosity equation:

$$B_i = H_i + \rho_i \sum_{j=1}^{N} F_{ij} B_j \qquad [4.28]$$

The term F_{ij} is the view factor between P_i and P_j elements; it is defined by:

$$F_{ij} = \frac{1}{A_i} \int_{A_j} \int_{A_i} \frac{\cos\theta_i \cos\theta_j}{\pi r^2} V(y_i, y_j)\, dA_i dA_j \qquad [4.29]$$

Discretization in a finite number N of elements thus reduces the problem of solving the integral radiosity equation [4.22] to that of solving a linear system of N equations with N unknowns Bj:

Radiosity equations express the balance of radiative exchanges between all possible pairs of elements of the scene, assuming diffuse reflections [SIL 94].

$$\begin{pmatrix} B_1 \\ B_2 \\ \vdots \\ B_n \end{pmatrix} = \begin{pmatrix} H_1 \\ H_2 \\ \vdots \\ H_n \end{pmatrix} + \begin{pmatrix} \rho_1 F_{11} & \rho_1 F_{12} & \cdots & \rho_1 F_{1n} \\ \rho_2 F_{21} & \rho_2 F_{22} & & \vdots \\ \vdots & & & \vdots \\ \rho_n F_{n1} & \cdots & \cdots & \rho_n F_{nn} \end{pmatrix} \begin{pmatrix} B_1 \\ B_2 \\ \vdots \\ B_n \end{pmatrix} \qquad [4.30]$$

In the above system, the components of the matrix are the reflection coefficients ρ_i and the view factors F_{ij}, also known as form factors [4.29]. The view factor is a purely geometrical variable, but its definition relies on energy concepts: F_{ij} is the proportion of the total power leaving P_i which is received by P_j. Therefore, it links two patches: P_i and P_j (Figure 4.2). The radiosities B_i and the exitances E_i are measured in W m^{-2}, while the view factors are dimensionless.

The solution of this system is easy because the matrix is very well conditioned. Many journals and books explain how to manage this problem [KAJ 86, COH 93, SIL 94, BEC 13a].

4.2.5. *Properties of the radiosity matrix*

The system of linear equations [4.30] works with the unknown radiosity B_i gathered in the vector B and the second member of exitances E_i, in the vector E. It is written in more compact form by introducing the radiosity matrix *R*.

$$RB = H; \quad R_{ij} = \delta_{ij} - \rho_i F_{ij} \qquad [4.31]$$

The view factor of a plane patch toward itself is equal to zero. As a consequence, in a closed cavity, the sum of the view factors of the elements surrounding a plane patch is equal to 1. This is the closure property.

$$\sum_{j=1}^{n} F_{ij} = 1 \quad ; \quad i = 1, n \qquad [4.32]$$

This property is verified by simple inspection of the view factors matrix: the sum of the terms of each row must be equal to 1. It can also be checked by solving the system of equations. If all the reflection coefficients are identical (all the walls have the same properties) and if the second member is composed of exitances all equal to 1, all the radiosities are identical:

$$E_i = 1 \quad ; \quad i = 1, n \quad \rightarrow \quad B_i = \frac{1}{1-\rho} \quad ; \quad i = 1, n \qquad [4.33]$$

The discrepancies with respect to [4.33] reflect the closure defaults of the view factors. Considering the symmetry of the integral [4.29], we also deduce the property of reciprocity. If A_i and A_j are the areas of the elements i and j, we obtain:

$$A_i F_{ij} = A_j F_{ji} \quad ; \quad F_{ji} = F_{ij} \; A_i / A_j \qquad [4.34]$$

This property means that, for instance, if A_i is a disk and A_j is the hemisphere above it, $F_{ij} = 1$ according to the closure condition [4.32]. Moreover, because the area of the hemisphere is twice that of the base disk, by [4.34], $F_{ji} = 1/2$. This means that the disk can only see the dome; meanwhile, the dome partly sees the disk and itself.

Like the previous one, this property can be verified by simple inspection of the view factors matrix. Note that if all elements have the same area (a condition seldom met), the matrix of view factors is symmetric.

The property of reciprocity can also be verified by solving the *equations of importance*. Denoting by A the vector (single column matrix) whose components are the areas of the elements, since radiosities are constant in each element, the power dissipated by the surfaces of the scene is calculated simply by the product:

$$p = B^T A = (R^{-1}H)^T A = H^T (R^{-1})^T A = H^T X \qquad [4.35]$$

The vector X defines the importance; it is the solution of the system of equations:

$$R^T X = A \qquad [4.36]$$

This system of equations, where the coefficient matrix is the transpose of [4.31], is the adjoint of the radiosity equations. These are the equations of importance. The quantities A and X are areas measured in m^2. Writing [4.36] explicitly:

$$\sum_{i=1}^{n} X_i R_{ij} = A_j \qquad [4.37]$$

Assuming that all the reflection coefficients are equal:

$$\sum_{i=1}^{n} X_i \left(\delta_{ij} - \rho F_{ij}\right) = A_j \qquad [4.38]$$

Hence:

$$X_j - \rho \sum_{i=1}^{n} X_i F_{ij} = A_j \quad \rightarrow \quad \frac{X_j}{A_j} - \rho \sum_{i=1}^{n} X_i \frac{F_{ij}}{A_j} = 1 \qquad [4.39]$$

and taking into account the reciprocity equation [4.34]:

$$\frac{X_j}{A_j} - \rho \sum_{i=1}^{n} X_i \frac{F_{ji}}{A_i} = 1 \qquad [4.40]$$

This equation is verified by the solution:

$$\frac{X_i}{A_i} = \frac{1}{1-\rho} \quad ; \; i = 1,\ldots,n \qquad [4.41]$$

If all the reflection coefficients are equal, the importances per areas are also equal. The discrepancies between the actual solution of [4.36] and the result [4.41] show the evidence of the deficiencies caused by the failure of the reciprocity equations.

According to [4.35], the power can be calculated in two ways:

$$p = B^T A = H^T X \qquad [4.42]$$

Let w be the power emitted by all sources.

$$w = H^T A \qquad [4.43]$$

In a closed room, if all the reflection coefficients are equal, a simple physical reasoning allows us to deduce the exact theoretical power developed by the system [BEC 11a].

$$p_{exact} = \frac{w}{1-\rho} = \frac{H^T A}{1-\rho} \qquad [4.44]$$

Power [4.44] is an input of the problem: it depends on exitances and areas which are, respectively, the second members of the radiosity and importance equations. The denominator depends only on the unique reflectance of the elements.

The solution of the radiosity equations gives the same result if the view factors are calculated exactly. However, since it is usually not possible in 3D, the comparison of the radiosity solution and the result [4.44] gives a useful indication of the error in calculating the view factors. We deduce the accuracy of the radiated power calculated from the radiosity or importance:

$$\frac{p}{p_{exact}} = \frac{(1-\rho)}{H^T A} B^T A = (1-\rho) \frac{H^T X}{H^T A} \qquad [4.45]$$

We now consider the n different orders of reflections. For the first one, we obtain ρE, with E, the total exitance. For the second one, we obtain $\rho^2 E$, and then $\rho^3 E$, etc. When the equilibrium is achieved, the total power P emitted by all the walls and boundaries is equal to the sum of the initial exitance and the infinite number of reflections [SUM 92, KAJ 86].

$$H_T = \left(1+\rho+\rho^2+\rho^3+...\right)H \qquad [4.46]$$

The limit of the previous series leads to a simple formula:

$$\forall \rho \in [0,1] \ , \ \sum_{n=0}^{\infty} \rho^n = \frac{1}{1-\rho} \quad \rightarrow \quad H_T = \frac{H}{1-\rho} \qquad [4.47]$$

This relation is valid if the domain is closed and if all the reflection coefficients are the same.

A typical application of these results is the use of the integrating sphere system to measure the diffuse reflectance and transmittance of materials [HAN 01]. These relations are linked to the fundamental properties of this equation which is a Fredholm integral of the second kind. The expression of the solution as a Neumann series also expresses the solution as an initial radiance followed by different reflections. It also explains why the solution of this equation can benefit from standard iterative methods and is very robust.

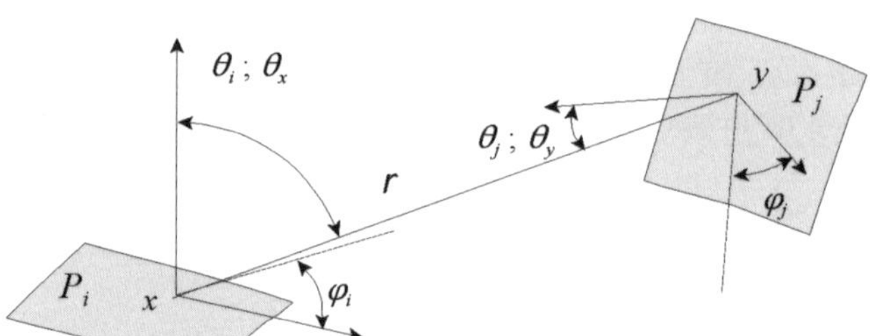

Figure 4.2. *View factor definition scheme*

4.3. View factors

The view factor is a pure geometric quantity, even if its definition is based on energetic theory. It specifies the fraction of energy leaving a surface Q_i that reaches another surface Q_j [COH 93].

$$F_{ij} = \frac{1}{A_i} \int_{x \in Q_i} \int_{y \in Q_j} \frac{\cos\theta_x \cos\theta_y}{\pi r^2} V_{ij}(x,y)\, dydx \qquad [4.48]$$

The factor F_{ij} connects two patches: Q_i with area A_i and Q_j with area A_j (Figure 4.2). The angle between the ray r and the normal to patch Q_i is denoted by θ_x while the angle between the ray and the normal to patch Q_j is denoted by θ_y. The symmetry of this integral gives way to the property of reciprocity [LAM 60]:

$$A_i F_{ij} = A_j F_{ji} \qquad [4.49]$$

The coefficient V_{ij} (x, y) is the visibility function between the points x and y located on Q_i and Q_j; it takes only the value 1 (visible) or 0 (occluded).

Closed solutions of the view factor exist for a few particular configurations [HOW 10], and, except for these special 3D configurations, it is not possible to calculate them accurately. We must therefore return to numerical methods.

One approach is to calculate it in two steps. First, we calculate the "differential view factor", "point wise factor" or, "point to area view factor" which is the inner core of expression [4.48] (Figure 4.3):

$$F_{x,Q_j} = \int_{y \in Q_j} \frac{\cos\theta_x \cos\theta_y}{\pi r^2} V_{x,j}(y)\, dy = \frac{1}{\pi} \int_{y \in Q_j} \cos\theta_x \; V_{x,j}(y) d\omega \qquad [4.50]$$

Then, this quantity is integrated with the concerned element. Classical methods can be used such as the Gauss integration technique for quadrilaterals and Newton Cotes for triangles [BBE 12].

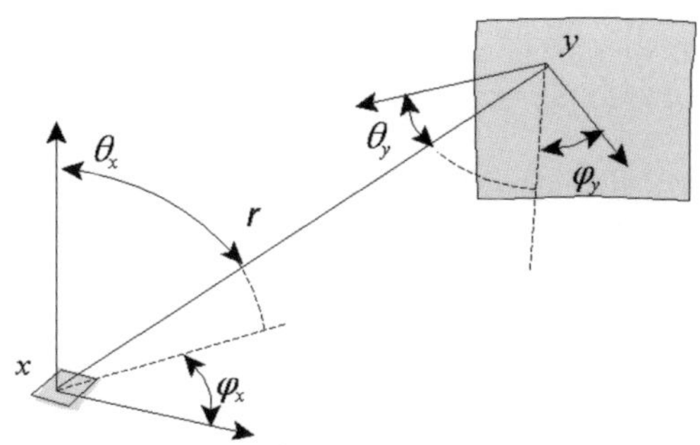

Figure 4.3. *View factor: point to patch – notations and general scheme*

In [4.50], dy is the differential area on patch Q_i and the variable $d\omega$ is the differential solid angle of patch Q_j. The integration is performed on the hemisphere built on the plane tangent to the patch Q_i at location x. By definition, if the patch Q_j covers the hemisphere, the differential view factor is equal to 1.

Computing the differential form in this way is rather simple, at least if it is not necessary to solve a visibility problem.

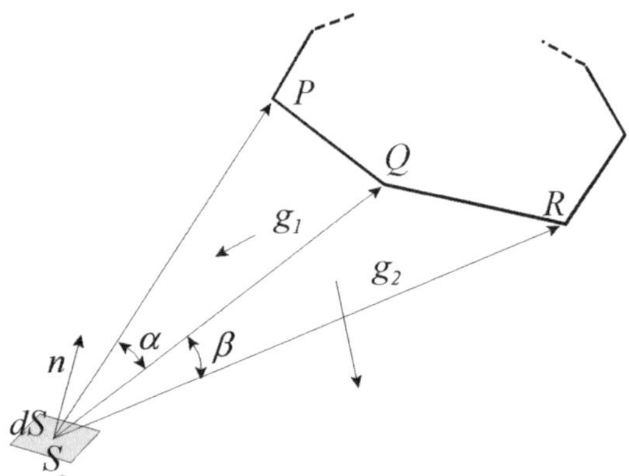

Figure 4.4. *View factor: point to patch – exact solution*

For a polygonal surface *P, Q, R*... (Figure 4.4), the following formula is exact if there are no obstacles [LAM 60, COH 93]:

$$F_{dS-j} = \frac{1}{2\pi}\sum_j n.g_j \qquad [4.51]$$

The vector n is normal to the surface supporting dS and on which we calculate the view factor. The vectors g_j are normal to the faces of the pyramid: *SPQ*, etc. Their modules are equal to the angles at the apex S of the pyramid: α, β, etc.

The other way to calculate the differential view factor is to project it onto the hemisphere defined at the concerned point and then to project the spherical polygon orthogonally on the plane tangent to the surface (the disk which is the base of the hemisphere). This projection is compared to the area of the disk. This interpretation of the view factor is known as the Nusselt analogy [NUS 28]. The calculation method is, in principle, easy to implement. Using an appropriate scaling, it is also possible to perform the computation of the areas on the stereographic projection instead of the orthogonal projection, with the advantage that the projections of spherical polygons are triangles whose edges are straight lines or circular arcs [BMB 11b].

If the patches Q_j seen from Q_i do not cover the full hemisphere, the complement to 2π of the sum of their view factors is called the sky view factor (closure property of the form factors [BEC 09]). The sky view factor is linked to the visible part of the vault of heaven; it is often used as a design parameter in architectural applications [NG 03, TAY 95].

4.4. Ray tracing

Ray tracing has been used since the 1970s for research in realistic rendering [SHI 03]. It is also used in the field of acoustics to simulate the behavior of large enclosed spaces in high frequencies.

Its principle is to send rays in different directions to detect their intersection with the elements of the scene and to deduce some properties. In practice, the main problem is to obtain a spatial distribution of directions as uniform as possible. We will examine several configurations: rays guided by a sphere, a hemisphere and a 90° sector (a quarter of a sphere).

The first question is whether a deterministic or random ray trace is performed. We will choose the random one because it eliminates aliasing problems. In principle, the deterministic ray tracing is more accurate, but in some situations it can over- or underestimate the number of impacts due to the position of the separations between geometrical elements.

In random ray tracing, it is difficult to obtain a uniform solution when defining the directions directly on the sphere or the other proposed geometries. This is why it is preferable to use a layered ray tracing called *stratified sampling*, which consists of creating a mesh and then randomly sending one or several rays inside the elements or cells.

The first issue is the choice of the mesh. We saw in Chapter 3 that the most appropriate one is based on elements of form as close as possible to the square. For this reason, we choose to work only on elements delimited by arcs of great or small circles. The small circles are orthogonal to the great circles. This type of mesh is very regular and suitable for random ray tracing (Figure 4.5).

The generation of quadrilateral elements on the sphere has been widely developed in [BEC 12a]. In this paper, we explain the rules to be followed to satisfy the condition of equal areas. The easiest method to generate equal-area cells is working on a planar projection of the sphere or hemisphere. Among these projections, the best candidates are Lambert equivalent projections (azimuthal and on the cylinder) or orthogonal azimuthal projection. In the latter case, under Nusselt analogy, the areas of the cells carried on the sphere will not be equal, but their view factors will.

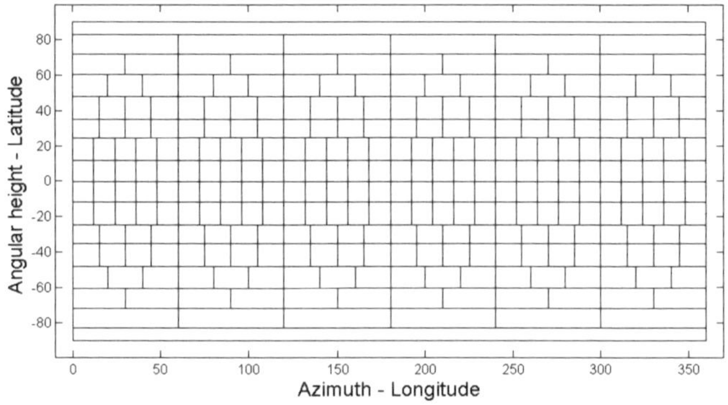

Figure 4.5. *Sphere tiling composed of 290 cells*

4.4.1. *Mesh quality*

In this mesh definition, there is one degree of freedom. We choose to respect as close as possible an aspect ratio equal to 1, either on a projection of the sphere or hemisphere or directly on the sphere.

As a first example of this method, we chose a mesh composed of 290 elements, 145 per hemisphere. It is well known because Tregenza [TRE 87] has proposed a version for the sky scanners. His version does not use equal-area elements, but a mesh in which he can define equal-area circles ensuring near-optimal coverage of the hemisphere. We give, hereinafter, a version with equal areas in the projection "plate carrée", which allows us to fully appreciate the high symmetry structure. The sphere is composed of two equal hemispheres roofed with polar caps, each hemisphere being divided into six equal parts.

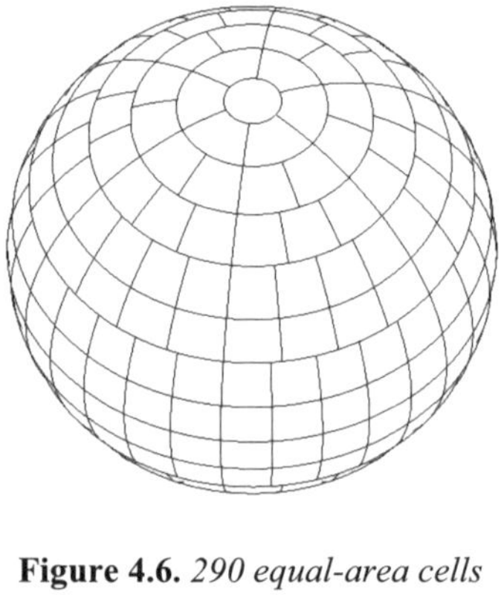

Figure 4.6. *290 equal-area cells*

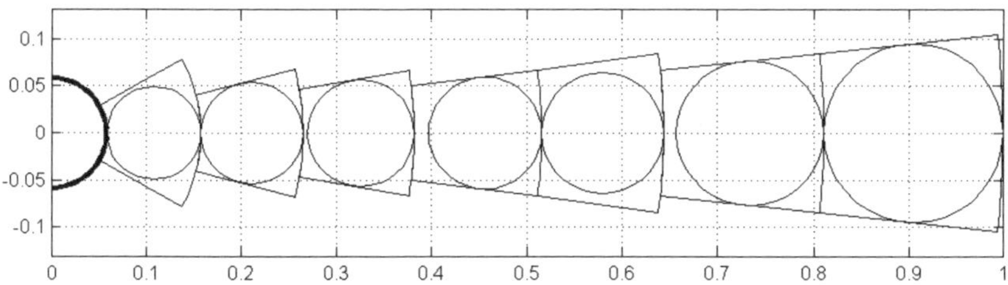

Figure 4.7. *Inscribed circles in a sector of the top hemisphere drawn in stereographic projection*

On the same model, we can use the full sphere or a part of it, if the polar cap cells are adapted accordingly.

A good method to control the quality of the tiling is to measure the coverage index of the mesh [TRE 87, BBE 13]. To compute it, we draw the greatest possible circle in all the cells. Multiplying the area of the smallest one by the number of cells and dividing this area by the hemisphere area gives the coverage index. It must be as close as possible to the maximum possible value of $\pi/4 = 0.7854$ that corresponds to its value for a disk inscribed in a square.

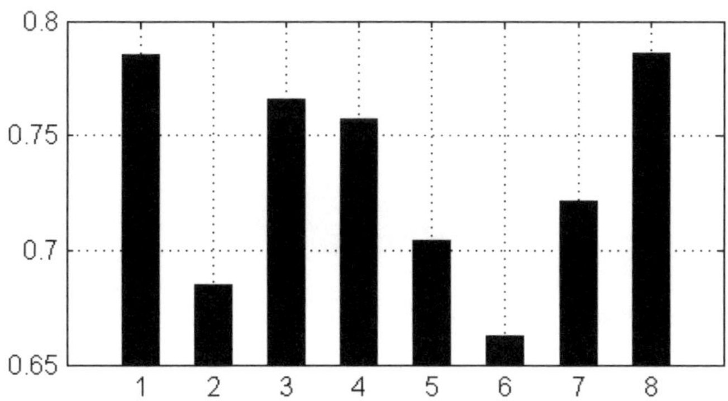

Figure 4.8. *Statistics of coverage index in the upper part of the 290 cells sphere*

The quality of the aspect ratio is assessed visually by examining whether the inscribed circles touch the four borders of the cells. In Figure 4.7, the situation is not perfect in rings 2 and 6. This observation is supported by the diagram of Figure 4.8. The center circle is counted as a ring and the coverage index is imposed on the maximum value obtained in a square.

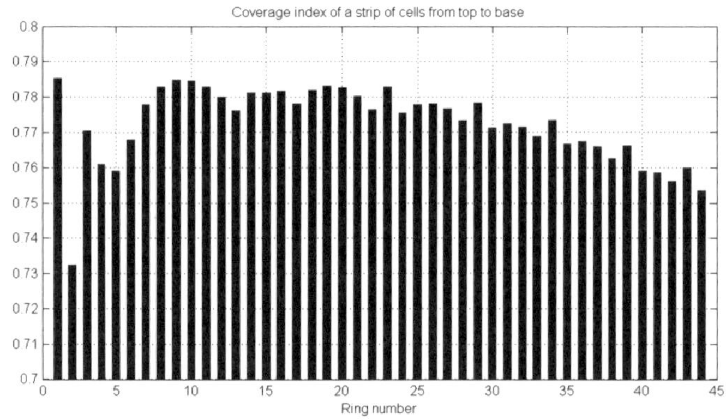

Figure 4.9. *Coverage index of the 44 rings of the 4,894 cells dome*

In practice, however, we choose much finer meshes, up to 20,000 per hemisphere if we want to precisely model the sky vault [BBE 13]. In another example, we have a fairly detailed mesh hemisphere. It contains nearly 5,000 items.

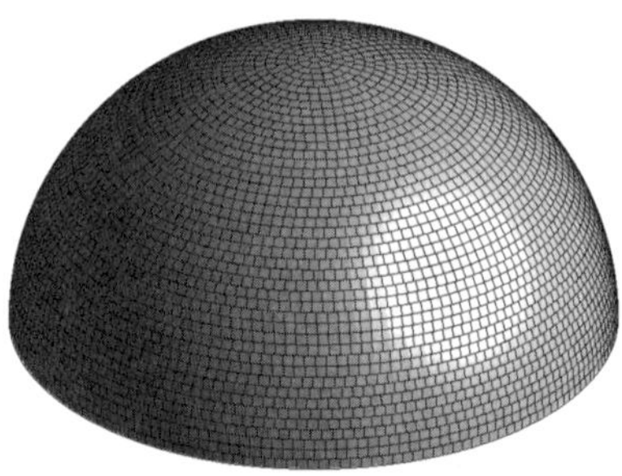

Figure 4.10. *Dome composed of 4,894 elements, Ic = 0.7325*

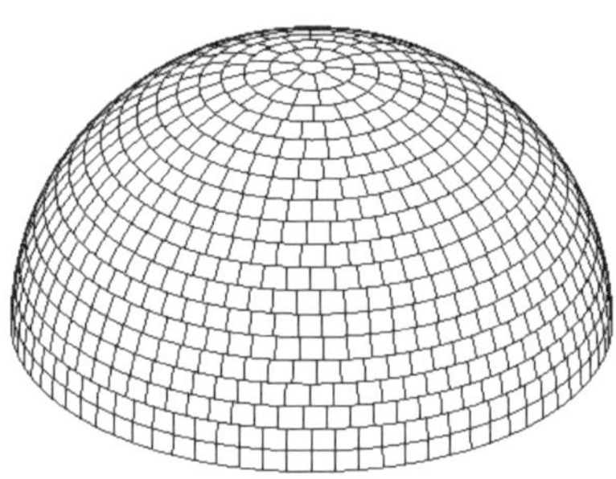

Figure 4.11. *Dome composed of 1,000 elements, Ic = 0.6489*

In this example, the statistics of coverage index is satisfactory, its minimum value being 0.73 in the ring following the central disk. When moving toward the base of the dome, the index is greater than 0.75, confirming that the cells are almost "square".

Having this aspect ratio close to one on the sphere or a coverage index around 0.7 ensures the uniformity of ray tracing. This is the option that will be chosen in the next tests (Figure 4.12).

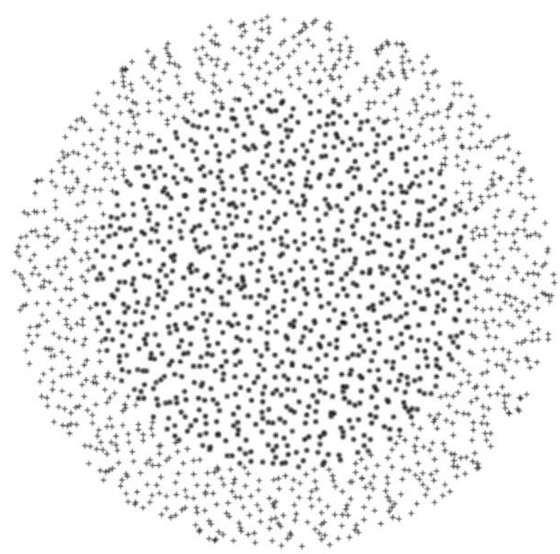

Figure 4.12. *Lambert projection of 2,000 random rays*

A simpler example of 1,000 cells was chosen. The coverage index seems poor, but its value is lower only in the first ring around the central disk. This is because the requirement was to obtain exactly 1,000 cells and to progress in the construction of the mesh from the outside to the inside. The cells are distributed in the rings as described in the vector below, where the first term represents the central circle, the second term gives the total number of cells in the central disk and the first ring: 9 additional cells. By adding 15 cells in the next ring, the total of 25 cells is obtained. This is the third term of the vector. The last ring has 79 cells and it leads to the total of 1,000 cells.

[1 10 25 45 72 105 143 186 234 287 345 407 473 542 614 688 764 842 921 1000]

In the Lambert projection of the deterministic ray tracing, the sphere is viewed from above; the spherical caps are located at the extremities of the vertical axis. To see the separation meridian of the mesh located on the rear part of the sphere (on the upper part of the drawing), it is convenient to impose a rotation of 90° to the east (Figure 4.13). This rotation is not performed for the representation of Figure 4.15.

Figure 4.13. *Lambert projection of 2,000 deterministic rays*

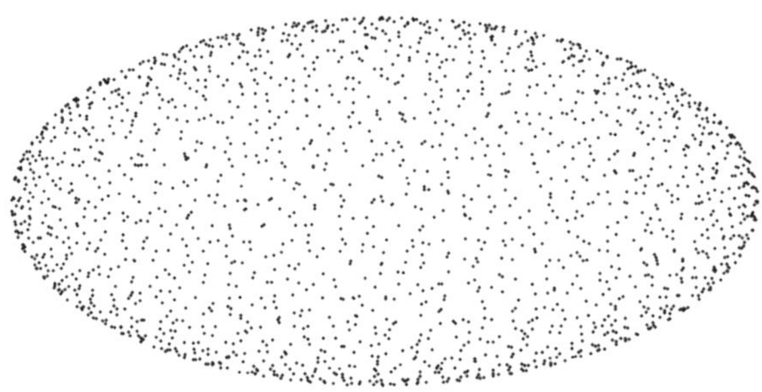

Figure 4.14. *Hammer projection of the full sphere composed of 2,000 random rays*

The representation of the thousand rays of this calculation is correct, but not perfect as we saw in Chapter 3 that, near the periphery, the cells are something distorted. To obtain a more uniform distribution on the projection, it is convenient to use the Hammer projection (Figure 4.14). This projection, presented in the previous section, also preserves the cell areas, but it deforms less than the Lambert projection on the ellipse periphery.

Another way to check the results is to visually compare the images of the same sphere in two random positions (Figures 4.16 and 4.17).

For a more thorough analysis of the sky radiation, we propose a tiling of heaven that consists of 20,000 elements [BBE 13], which corresponds to a grid of nearly 1°.

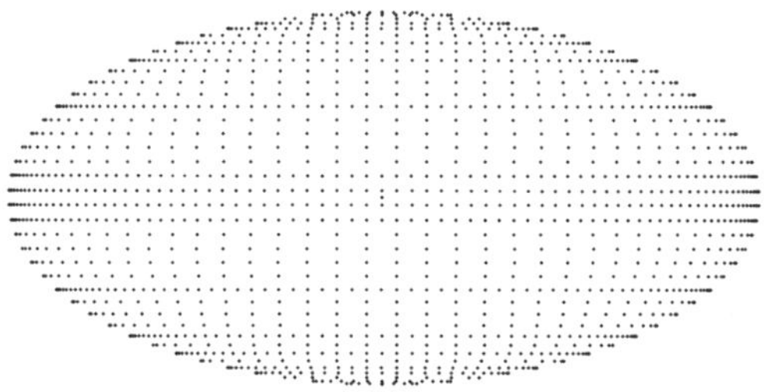

Figure 4.15. *Hammer projection of the full sphere – 2,000 deterministic rays*

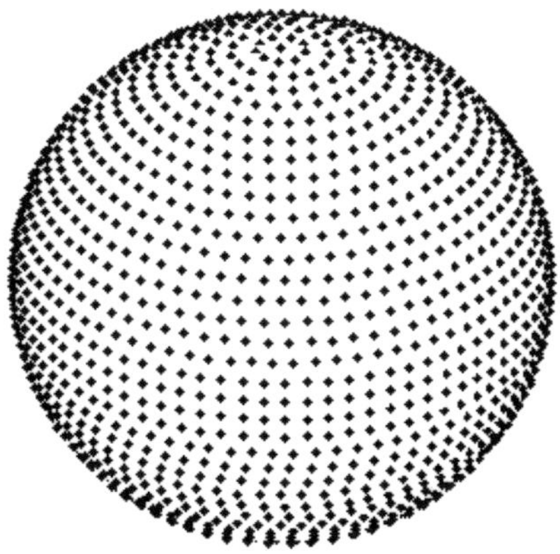

Figure 4.16. *Side view of the sphere with 2,000 rays*

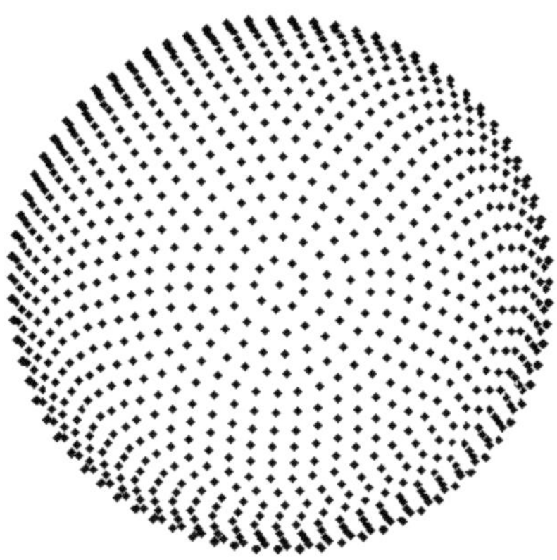

Figure 4.17. *Top view of the sphere – 2,000 rays*

The method just described is intended to create on the sphere or hemisphere equal-area cells of aspect ratio close to 1. These cells describe constant and compact solid angles on the spherical element. As they are quadrilaterals, it is particularly easy to use Monte Carlo ray tracing with stratified sampling, where the rays seek one or more random positions in each cell.

On all the equivalent projections of the preceding illustrations, it is clear that, in the case of the random ray tracing, the rays are uniformly distributed.

Ray tracing itself is a very simple technique when used in the context of the proposed mesh types, i.e. meshes defined between two parallels and two meridians. It follows that restricting the ray generation to a portion of the sphere itself comprised between two meridians and two parallels is also a trivial operation.

To achieve random shooting, we simply generate two sequences of random numbers between 0 and 1. These sequences will ideally have a dimension equal to the number of cells. When examining the cell *i*, the value *u(i)* of the first random number and the value *v(i)* of the second series are selected. Then, the orientation of the ray is simply given by its latitude and longitude and is calculated as follows:

$$\begin{aligned} \alpha(i) &= \begin{array}{l} \alpha(i)_{left}\left(1-u(i)\right)+\alpha(i)_{right}\,u(i); \\ \alpha_{left} \leq \alpha(i) \leq \alpha_{right} \end{array} \\ \varphi(i) &= \begin{array}{l} \varphi(i)_{bottom}\left(1-v(i)\right)+\varphi(i)_{top}\,v(i); \\ \varphi_{bottom} \leq \varphi(i) \leq \varphi_{top} \end{array} \end{aligned} \qquad [4.52]$$

When we have the definition of cells, the sequence of directions is easy to generate and is well suited to develop any type of adaptation or variant.

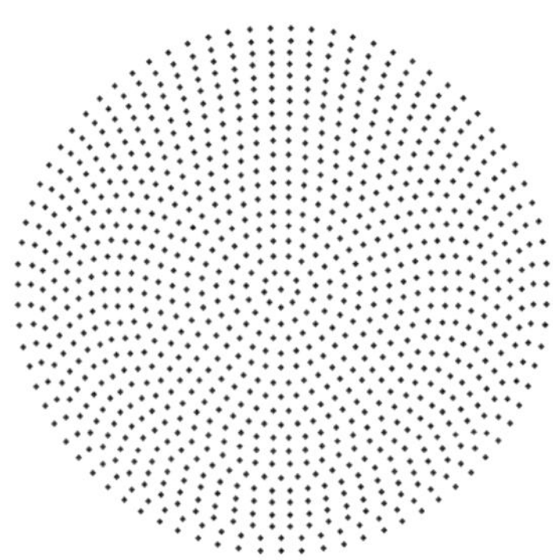

Figure 4.18. *Hemisphere with 1,000 rays, constant view factor cells*

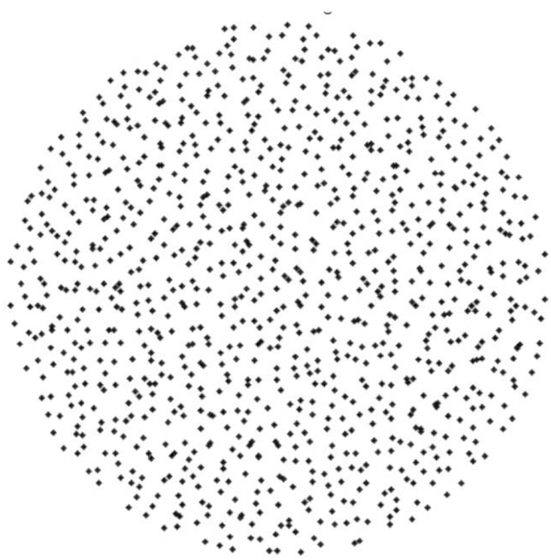

Figure 4.19. *Hemisphere with 1,000 rays, constant view factor cells, importance and stratified sampling*

4.4.2. *Solid angle or view factor*

In the computation of the solid angle, we take advantage of the possibility to use cells having the same area, which, in principle, allows us to accelerate the calculation. The calculation of view factor rather than solid angle is achieved by a method called *importance sampling technique*, where the areas of the elements assigned to a ray are multiplied by the cosine of the complement of the zenithal distance of the local hemisphere. Let us observe that the view factor is computed in relation to at least one surface. This means that the ray generation is performed with respect to a hemisphere or a smaller part of the sphere if the point is located at the intersection of two or more surfaces.

Based on the Nusselt analogy, we can also create cells that do not have the same area, but the same view factor. For this purpose, it is sufficient, as in the previous method, to create equal area cells on the unit disk, and then switch back to the sphere by the inverse law of that which was used to calculate the orthogonal azimuthal projection.

In subsequent operations, as all cast rays correspond to the same view factor, counting them on a certain area and dividing by the total number of cells will automatically provide the view factor of the area as a percentage of the total view factor.

4.5. Specular reflection of light and sound

Sensory geometry works with notions sometimes known since antiquity, or from the 17th Century or, more often, since the 19th Century, when almost all the necessary theories were completed and systematized. The 20th Century has brought essentially new algorithmic methods (Fourier transforms, radiosity methods, Monte-Carlo integral, etc.), while their large physical models have consciously departed from the sensory world. The new algorithms, and the computer revolution behind them, now allow us to recover the old ideas, which had been discarded as annoying or for the notion of merely being impossible to perform manually.

We show an example in the field of room acoustics. Let us first recall the definition of the sound level, which is the parameter studied here.

The acoustic energy can be expressed as a *power* (W, watts) as an *intensity* (I, energy per unit area, in joules per square meter) or by the square of the *rms pressure* (P_{rms} in pascals). Without going into detail, we assume here the relationship between the latter two variables for the progressive waves:

$$I \approx \frac{p_{rms}^2}{400} \qquad [4.53]$$

At a frequency of 1,000 Hz, the thresholds of perception are rms pressures 2.10^{-5} Pa (threshold of human hearing) and 20 Pa (threshold of pain). The first serves as a reference to set the sound pressure level, which is consequently equal to 0 and 120 dB for the two thresholds, respectively:

$$L_p = 10 \log \frac{p_{rms}^2}{p_0^2} = 20 \log \frac{p_{rms}}{p_0} \qquad [4.54]$$

A similar expression is the intensity level, whose reference I_0 is derived from P_0 and from equation [4.53]:

$$L_I = 10 \log \frac{I}{I_0} \quad \text{where} \quad I_0 = 10^{-12} J/m^2 \qquad [4.55]$$

In practice, these two terms are interchangeable, and we can use either sound pressure level (L_p) or intensity level (L_I).

A sound source is described by its power (in watts) or by its power level:

$$L_W = 10\log\frac{W}{W_0} \quad \text{where} \quad W_0 = 10^{-12}W \qquad [4.56]$$

The following table shows the levels achieved by different sound sources:

	W (watts)	L_W (dB)
female voice male voice	0.002 0.004	93 96
clarinet	0.05	107
violin	0.16	112
orchestra	$10 \Rightarrow 70$	$130 \Rightarrow 138$

In general, sound source emission is not omnidirectional: the body of the actor or the musician, for example, is an obstacle, and it limits the backward propagation. It is observed that the attenuation at high frequencies is high (it can exceed 15 dB), but almost inexistent at low frequencies, due to diffraction.

In any case, building acoustics still considers these phenomena as simple corrections to the standard case of the spherical emission (omnidirectional), which is characterized by a decrease in sound intensity proportional to the square of the distance (the emitted power is spread over an ever larger sphere, as the sound moves away from the point of emission):

$$I = \frac{W}{4\pi r^2} \qquad [4.57]$$

Previous equations then combine in:

$$L_I = L_W - 20\log r - 11 \qquad [4.58]$$

This formula indicates that if a source emits 100 dB, for example, it will generate a sound level of 89 dB at 1 m distance. Then, L_I will decrease by 6 dB for each doubling of distance: it will thus be 83 dB at 2 m, 77 dB at 4 m, 71 dB at 8 m, 65 dB at 16 m, etc.

To add two sound levels, we have to sum energies (intensities, powers, square rms pressure) and not levels. Thus, two contributions I_1 and I_2 give a total intensity $I = I_1 + I_2$, corresponding to a level:

$$L_I = 10\log\frac{I_1 + I_2}{I_0} = 10\log(10^{0.1L_1} + 10^{0.1L_2}) \qquad [4.59]$$

If both contributions are equal ($I_1 = I_2$), the resulting level increases by 3 dB:

$$L_I = 10\log\frac{2I_1}{I_0} = 10\log\frac{I_1}{I_0} + 10\log 2 = L_{I_1} + 3 \qquad [4.60]$$

Now consider a simple problem, well known by practitioners in room acoustics. It comes to finding the ideal form of an acoustic ceiling, which has to reflect the sound to the public so as to obtain the most uniform possible sound level. Indeed, the direct sound level decreases by 6 dB for each doubling of distance, and it may go below the noise level for the most distant spectators.

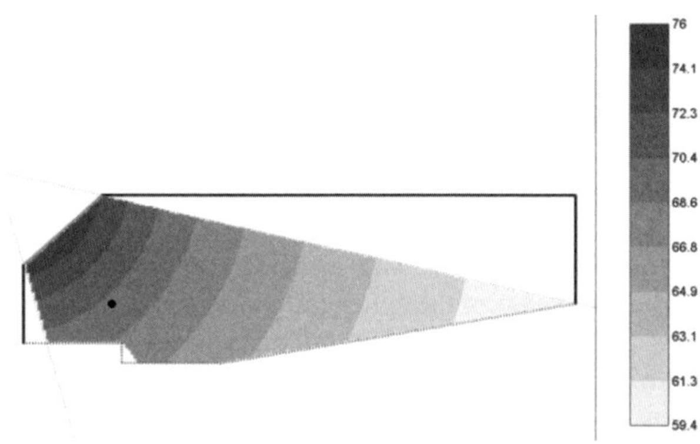

Figure 4.20. *Specular reflection*

Each flat section of the ceiling is like a mirror (specular reflection) to be oriented to send the reflected sound toward the back of the room. Figure 4.20 shows a section of the room (Radit2D software [BEC 02]). The floor of the stage and the tiers are considered to be absorbent. The source is located 1.5 m above the edge of the stage. We show here the first reflection off the first reflector directed toward all the public. The sound level (expressed in dB) decreases as the reflection moves away. Instead of tracing rays, the images method is used: each image is the symmetrical point of the source relative to each mirror.

One reflection is not sufficient. Indeed, the source image will always be farther from the public than the real source (i.e. the reflection path is necessarily longer than the direct path). So, a single reflection has very little effect. We will design the rest of the ceiling so as to multiply the number of reflections reaching the public. From the second reflector, it is no longer possible to cover the entire public. The rule is to cover at least the farthest public, since the closest spectators do not need reinforcement. Finally, Figure 4.21 shows, how, correctly oriented, each plane sector makes a discretized curve-shaped ceiling appear.

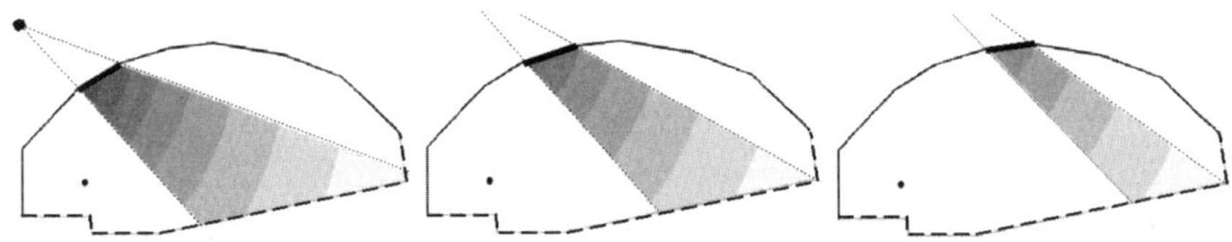

Figure 4.21. *Step-by-step acoustical design of the ceiling*

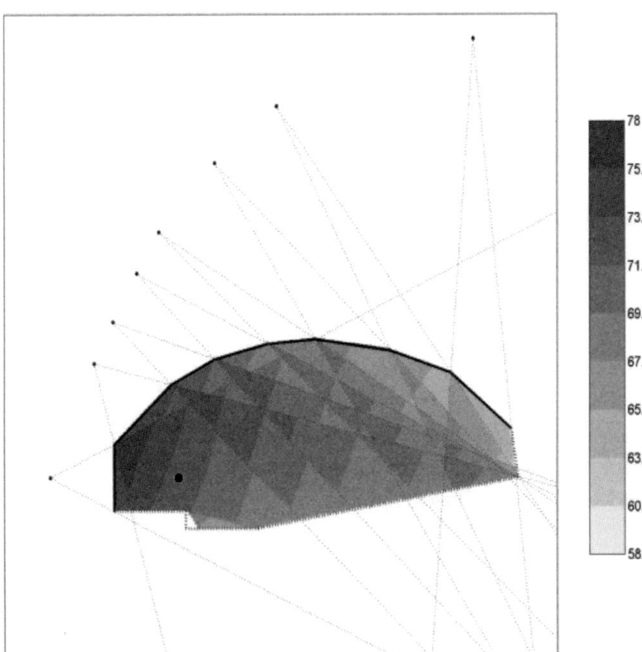

Figure 4.22. *Designed ceiling and orthotomic curve*

A pseudo-elliptical ceiling is obtained, which is typical in current rooms. What is it? Is this really an ellipse? In Figure 4.22, all images are shown. It is clear that if the discretization is finer, the ceiling becomes a curve, and the images build another curve, called orthotomic. Here, the orthotomic has the appearance of a circle. But is it really a circle? Figure 4.23 shows a well-known case: a source is located on one of the two focal points of an elliptical mirror. All rays from this source go to reflect on the second focal point. The images then form a perfect circle. When the orthotomic becomes circular, it is because there is a focalization. This is not what is wanted in the design of an acoustic ceiling: energy must be reflected to an area – the rear part of the room – but especially must not concentrate at one point. An elliptical ceiling is therefore not ideal, far from it.

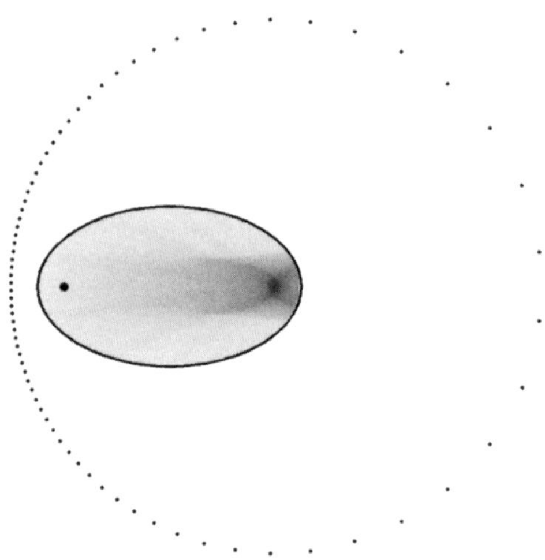

Figure 4.23. *Focalization by an ellipse*

The optimal solution of this problem is not known in geometrical acoustics. However, it has been found in a completely different field – geometrical optics – and more specifically by a lamp designer.

In the early 1920s, the designer Poul Henningsen proposed a new lamp, the result of years of research into lighting devices, with particular attention to geometry [RYB 00].

The problem was to compose a lighting system with several reflectors which do not interfere with each other and which ensure a good distribution of the light. The famous PH lamp, the result of this research, is built with three surfaces of revolution whose generatrix is a piece of logarithmic spiral.

Every curve defines, with respect to any point that we call "source", a series of curves accompanying it, and displaying its properties of reflection. Here, we are interested in two of these accompanying curves: the orthotomic, image of the curve with respect to the source, and the caustic, envelope of the rays reflected by the curve.

Without going into detail we recall that the logarithmic spiral function is an exponential so that it derives in itself, which explains the following property: the orthotomic and the caustic of reflection of an optical device in the form of a logarithmic spiral are also logarithmic spirals when the source occupies the apex of the device.

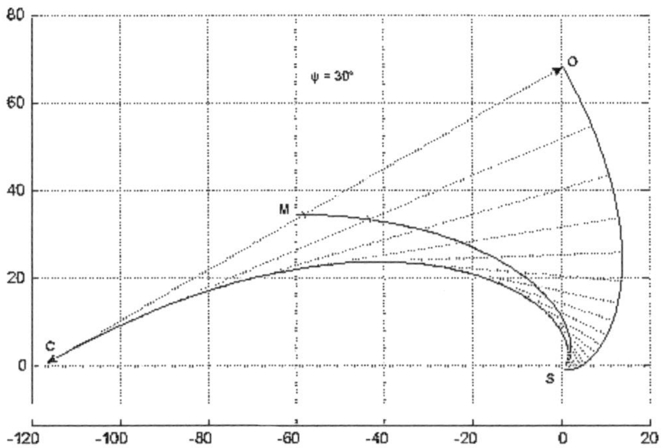

Figure 4.24. *Logarithmic spiral, its caustic and its orthotomic*

In Figure 4.24, the source *S* is at the apex of a section of logarithmic spiral that ends at the point M. Considering this section as a mirror, we find images of *S* distributed over the orthotomic ending in *O*. The rays reflected by the mirror are all tangent to the caustic ending in *C*. The three curves are sections of logarithmic spiral, originating in the same apex *S*.

The PH lamp benefits from this property. Indeed, if we place the light bulb at the apex of a surface of revolution built on the logarithmic spiral, we obtain the configuration of Figure 4.25: the entire light incident on the reflector (in the centre) is reflected within a beam comprised between the reflector and its caustic (below). We can then deploy a second reflector following this caustic: it will have the same shape as the first, and the light reflected by the first reflector will rub the second without having a second reflection in it.

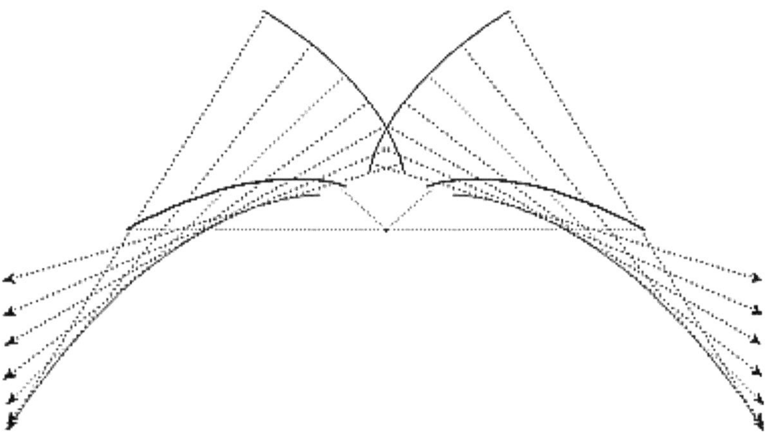

Figure 4.25. *PH luminaire design*

In summary, Poul Henningsen took two very remarkable properties of the logarithmic spiral: its ability to reproduce itself and the containment of its reflections on a beam limited by a caustic with the same shape. Both properties lead us to work this curve by sections, as in the PH lamp. We can build an acoustic device based on the same principle. Indeed, a roof consisting of logarithmic spiral sections can reinforce, through the reflections, the sound level at the rear part of a room, without affecting the part next to the source, which, to be in the direct field, does not need help. It is the shape we were looking for.

Conclusion

This book provides four main contributions in the field of geometry in its two aspects: perception of the 3D environment and the solution or simulation of practical engineering problems or physical situations.

The first aspect is concerned with the justification of the central projection, which is a consistent method to validate human behavior in order to understand and manage the 3D sensory space. This model was discovered six centuries ago and developed by painters and architects of the Renaissance. With the emergence of computer simulations in the field of representation, it has been receiving new attention, resulting in original enhancements, including new optical devices.

The second aspect focuses on the sphere. The study of spherical curves and polygons, just as the evaluation of lengths and areas on spherical objects, is the necessary constituent for the control of pointwise sources of radiation, 360° scenes representation and sphere entities. A synthesis of world cartography is also performed with the construction of circular, rectangular and elliptic maps. Their characteristics are explained in order to solve specific problems such as the evaluation of solid angles and view factors.

The third point consists of a brief presentation of the homogeneous coordinates. They provide the essential tools to manage objects in the 3D space. Their relation with the central projection helps us to understand the fundamental properties of the conical curves and how to control them with polynomial functions.

Finally, an original method is presented to mesh the sphere. It is supposed to provide, in the near future, very powerful and fast tools in the field of radiative transfers in light, heat and acoustic domains. Some suggestions are given for the solution of radiosity equations and for the use of important and stratified sampling in the techniques of Monte Carlo ray tracing.

However, this book is not just a collection of problems and solutions. We tried to show the inherent consistency of geometry, from its ancient birth in human perception – visual and musical – to its latest digital applications in different fields of radiation physics.

This choice allowed us to highlight the original properties shared by scientific areas that are still too fragmented, and to show both the unity and the variety of geometric ways of thinking.

For instance, conic curves are visually reduced to the circle, and numerically to the parabola.

The logarithmic spiral appears not only as a rhumb line in mapping applications, but also as a generatrix for optimal acoustic ceilings or luminaire reflectors.

Geometric constructions, such as the cross ratio and foreshortening, structure our different perceptions of the world, and give them unity, range and consistency.

Bibliography

[ARI 94] ARISTOTLE, *On the Soul*, SMITH J.A. (trans.), Book 3, part 2, The Internet Classics Archive by Daniel C. Stevenson, Web Atomics, 1994.

[ARV 95] ARVO J., "Stratified sampling of spherical triangles", *SIGGRAPH '95*, August 1995.

[AUP 93] AUPPERLE L., HANRAHAN P., "A hierarchical illumination algorithm for surfaces with glossy reflection", *SIGGRAPH '93, Proceedings of the 20th Annual Conference on Computer Graphics and Interactive Techniques*, ACM, New York, NY, pp. 155–162, 1993.

[BEC 91] BECKERS P., "Recent developments in shape sensitivity analysis: the physical approach", *Engineering Optimization*, vol. 18, nos. 1–3, pp. 67–78, 1991.

[BEC 02] BECKERS B., MASSET L., Radit2D, computer program, 2002. Available at www.heliodon.net. [Manual in Spanish]

[BEC 03] BECKERS P., *Communication Graphique*, University of Liège, Belgium, 2003.

[BEC 05] BECKERS B., Geometría sensible, PhD Thesis, UPC, Spain, 2005.

[BEC 09] BECKERS B., "Geometrical interpretation of sky light in architecture projects", *Actes de la Conférence Internationale Scientifique pour le BATiment CISBAT 2009*, EPFL, Lausanne, Switzerland, September 2009.

[BEC 11a] BECKERS B., "Impact of solar energy on cities sustainability", in BODART M., EVRARD A. (eds), *Architecture & Sustainable Development, Proceedings of 27th International Conference on Passive and Low Energy Architecture (PLEA '11)*, Louvain-la-Neuve, Belgium, vol. 2, pp. 395–400, 13–15 July 2011.

[BEC 11b] BECKERS B., MASSET L., BECKERS P., "The universal projection for computing data carried on the hemisphere", *Computer-Aided Design*, vol. 43, pp. 219–226, 2011.

[BEC 12a] BECKERS B., BECKERS P., "A general rule for disk and hemisphere partition into equal area cells", *Computational Geometry*, vol. 45, no. 7, pp. 275–283, 2012.

[BEC 12b] BECKERS B., BECKERS P., "Radiative simulation methods", in BECKERS B. (ed.), *Solar Energy at Urban Scale*, Chapter 10, ISTE, London and John Wiley & Sons, New York, pp. 205–236, 2012.

[BEC 13a] BECKERS B., "Taking advantage of low radiative coupling in 3D urban models", *Eurographics Workshop on Urban Data Modelling and Visualisation*, Girona, Spain, 6–10 May 2013.

[BEC 13b] BECKERS B., BECKERS P., "Sky vault partition for computing daylight availability and shortwave energy budget on an urban scale", *Lighting Research & Technology*, 6 September 2013.

[BES 05] BESUIEVSKY G., MARTIN I., "A hierarchical algorithm for radiosity daylighting", report, University of Gerona, 2005.

[BJO 03] BJØRKE J.T., GRYTTEN J.K., HÆGER M., *et al.*, "A global grid model based on 'constant area' quadrilaterals", *Proceedings of the 9th Scandinavian Research Conference on Geographical Information Science (ScanGIS '03)*, Espoo, Finland, pp. 239–250, 4–6 June 2003.

[BOR 86] BORCEUX F., *Invitation à la géométrie*, CIAO, Belgium, 1986.

[CHA 04] CHAIWIWATWORAKUL P., CHIRARATTANANON S., "Evaluation of sky luminance and radiance models using data of North Bangkok", *Leukos,* vol. 1, no. 2, pp. 107–126, October 2004.

[COH 85] COHEN M.F., GREENBERG D., "The hemicube: a radiosity solution for complex environments", *SIGGRAPH'85,* vol. 19, no. 3 pp. 31–40, 1985.

[COH 93] COHEN M.F., WALLACE J.R., *Radiosity and Realistic Image Synthesis*, Academic Press, 1993.

[CRI 00] CRITTENDEN R.G., "Igloo pixelizations of the sky", *Astrophysical letters & communications*, vol. 37, pp. 377–382, 2000.

[DAL 33] D'ALEXANDRIE P., *La collection*, Liber III, proposition VII-129, page 69, 98 of the BNF document. Pappus d'Alexandrie (trans. Paul ver Eecke), La Collection Mathématique de Pappus d'Alexandrie (reprint. 1982) (2 vol.), Paris, Libr. A. Blanchard, 1933.

[DU 03] DU Q., GUNZBURGER M.D., JU L., "Voronoï-based finite volume methods, optimal Voronoi meshes, and PDEs on the sphere", *Comput. Methods Appl. Mech. Engrg.*, vol. 192, pp. 3933–3957, 2003.

[EUC 00] EUCLIDES, *Óptica*, GARCÍA P.O. (trans.), Editorial Gredos, Spain, 2000.

[FAR 90] FARIN G., *Curves and Surfaces for Computer Aided Geometric Design: A Practical Guide*, 2nd ed., Academic Press, 1990.

[FAU 08] FAUCHEREAU N., SINCLAIR S., PEGRAM G., "2-D empirical mode AAecomposition on the sphere, application to the spatial scales of surface temperature variations", *Hydrology and Earth System Sciences Discussions*, vol. 5, pp. 405–435, 2008.

[FER 11] FERNANDEZ S., *Lobachevski: un espiritu indomable*, NIVOLA, Spain, 2011.

[FOL 90] FOLEY J.D., VAN DAM A., FEINER S.K., *et al.*, *Computer Graphics, Principles and Practice*, 2nd ed., Addison-Wesley Publishing Company, 1990.

[FRA 72] FRAEIJS DE VEUBEKE B., SANDER G., BECKERS P., Dual analysis by finite elements. Linear and non linear applications, Technical Report AFFDL-TR-72-93, Wright-Patterson Air Force Base, 1972.

[FRA 84] FRANCESCA D.P., *De prospectiva pigendi*, italian version by Giusta Nicco-Fasola, Casa Editrice Le Lettere, Florence, 1984.

[FRE 08] FREY P.J, GEORGE P.L., *Mesh Generation Application to Finite Elements*, 2nd edition, ISTE, London, & John Wiley & Sons, New York, 2008.

[GAS 08] GASCUEL J.-D., HOLZSCHUCH N., FOURNIER G., *et al.*, "Fast non-ninear projections using graphics hardware", *ACM Symposium on Interactive 3D Graphics and Games*, 2008.

[GEO 98] GEORGE P.L., BOROUCHAKI H., *Delaunay Triangulation and Meshing: Application to Finite Elements*, Hermes, Paris, 1998.

[GON 09] GONZÁLES A., "Measurement of areas on a sphere using Fibonacci and latitude longitude lattices", *Mathematical Geosciences*, vol. 42, pp. 49–64, 2010.

[GRE 86] GREENBERG D.P., COHEN M.F., TORRANCE K.E., "Radiosity: a method for computing global illumination", *The Visual Computer*, vol. 2, pp. 291–297, 1986.

[GRI 92] GRINGORTEN I.I., YEPEZ P.J., The division of a circle or spherical surface into equal-area cells or pixels, Phillips Laboratory Report 92-27740, 1992.

[HAN 01] HANSSEN L., "Integrating-sphere system and method for absolute measurement of transmittance, reflectance, and absorptance of specular samples", *Applied Optics*, vol. 40, no 19/1, pp. 3196–3204, July 2001.

[HEC 92] HECKBERT P.S., "Introduction to global illumination", *SIGGRAPH '92*, 1992.

[HEC 93] HECKBERT P.S., "Finite element methods for radiosity", appeared in *Global Illumination Course Notes, SIGGRAPH '93*, Anaheim, August 1993.

[HOW 10] HOWELL J.R., SIEGEL R., MENGUC M.P., *Thermal Radiation Heat Transfer*, 5th ed., Taylor & Francis/CRC, New York, 2010. [Reference is also made to: A catalog of radiation heat transfer configuration factors, http://www.engr.uky.edu/rtl/Catalog/.]

[KAJ 86] KAJIYA J.T., "The rendering equation", *SIGGRAPH '86*, Dallas, vol. 20, no. 4, 18–22 August 1986.

[LAM 60] LAMBERT J.H., *Photometria sive de mensura et gradibus luminis, colorum et umbrae*, Augsburg, C. Detleffsen for the widow of Eberhard Klett, 1760.

[LAP 11] LAPAINE M., "Mollweide map projection", *KoG*-15, pp. 7–16, 2011.

[LAV 13] LAVALLOU F., "Un invariant fondamental: le birapport. Invariance par dualité. D'où la photo a-t-elle été prise?", *Tangente hors série no. 47: La magie des invariants*, Editions Pole, pp. 88–101, 2013.

[LEI 55] LEIGHLY J.B., "Aitoff and hammer: an attempt at clarification", *Geographical Review*, vol. 45, no. 2, pp. 246–249, April 1955.

[LEO 06] LEOPARDI P., "A partition of the unit sphere into regions of equal area and small diameter", *Electronic Transactions on Numerical Analysis*, vol. 25, pp. 309–327, 2006.

[LES 90] LESSING G.E., *Laocoonte*, Berlin, 1766, Spanish version of Eustaquio Barjau, Editorial Tecnos, Madrid, 1990.

[LIE 01] LIENARD P., *Petite histoire de l'acoustique, Bruits, sons et musique*, Hermès, Lavoisier, 2001.

[MAR 13] MARQUES R., BOUVILLE C., RIBARDIÈRE M., *et al.*, "Spherical fibonacci point sets for illumination integrals", *Computer Graphics Forum*, vol. 32, no. 8, pp. 134–143, December 2013.

[MEU 10] MEULDERS M., *Helmholtz: From Enlightenment to Neuroscience*, GAREY L. (trans.), MIT Press, 2010.

[MIG 00] MIGUET F., Paramètres physiques des ambiances architecturales: Un modèle numérique pour la simulation de la lumière naturelle dans le projet urbain, PhD Thesis, Ecole Polytechnique de l'Université de Nantes, Ecole d'Architecture de Nantes, 2000.

[MIT 96] MITCHELL D.P., "Consequences of stratified sampling in graphics", *SIGGRAPH '96*, pp. 277–280, August 1996.

[NEE 97] NEEDHAM T., *Visual Complex Analysis*, Oxford University Press, New York, 1997.

[NG 03] NG E., "Applying computational simulation results to the development of a design method for daylighting design and regulation in high-density cities", *8th International IBPSA Conference*, Eindhoven, Netherlands, 11–14 August 2003.

[NIS 86] NISHITA T., NAKAMAE E., "Continuous tone representation of three-dimensional objects illuminated by sky light", *ACM Transactions on Graphics*, vol. 20, no. 4, pp. 125–132, 1986.

[NON 93] NONO L., *Écrits*, French version supervised by Laurent Feneyrou, BOURGOIS C. (ed.), 1993.

[NUS 28] NUSSELT W., "Graphische bestimmung des winkelverhaltnisses bei der wärmestrahlung", *Zeitschrift des Vereines Deutscher Ingenieure,* vol. 72, no. 20, p. 673, 1928; see BECKERS B., MASSET L., BECKERS P., "Commentaires sur l'analogie de Nusselt", Report Helio_004_fr, 2009, available at http://www.heliodon.net/.

[PAN 75] PANOFSKY E., *La perspective comme forme symbolique*, Editions de Minuit, 1975.

[PRO 56] PROUHET E., "Note sur l'aire du triangle sphérique, formule de Lhuilier", *Nouvelles Annales de Mathématiques*, 1st series, vol. 15, pp. 91–93, 1856.

[RAS 03] RASSINEUX A., BREITKOPF P., VILLON P., "Simultaneous surface and tetrahedron mesh adaptation using mesh-free techniques", *International Journal for Numerical Methods in Engineering*, vol. 57, pp. 371–389, 2003.

[RYB 00] RYBERG B., *Light Years Ahead: The Story of the PH Lamp*, Copenhagen University, 2000.

[ROS 04] ROSELL V., VIVÓ R., "Improving form factor accuracy by projecting onto the hemisphere base", *WSCG Posters Proceedings*, 2004.

[SAF 97] SAFF E.B., KUIJLAARS A.B.J., "Distributing many points on a sphere", *The Mathematical Intelligencer*, vol. 19, no. 1, 1997.

[SAH 98] SAHR K., WHITE D., "Discrete global grid systems", *Computing Science and Statistics*, vol. 30, pp. 269–278, 1998.

[SAH 08] SAHR K., "Location coding on icosahedral aperture 3 hexagon discrete global grids", *Computers, Environment and Urban Systems*, vol. 32, pp. 174–187, 2008.

[SHA 88] SHAO M.-Z., PENG Q.-S., LIANG Y.-D., "A new radiosity approach by procedural refinement for realistic image synthesis", *Computer Graphics*, vol. 22, no. 4, August 1988.

[SHA 96] SHAO M.-Z., BADLER N.I., Spherical sampling by archimedes' theorem, Technical Report MS-CIS-96-02, University of Pennsylvania, 1996.

[SHI 97a] SHIRLEY P., CHIU K., "A low distortion map between disk and square", *Journal of Graphics Tools*, vol 2, no 3, pp. 45–52, 1997.

[SHI 97b] SHIRLEY P., MORLEY R.K., *Realistic Ray Tracing*, 2nd ed., A K Peters/CRC Press, 9 July 2003.

[SIL 89] SILLION F., PUECH C., "A general two-pass method integrating specular and diffuse reflection", *Computer Graphics*, vol. 23, no. 3, 1989.

[SIL 94] SILLION F., PUECH C., *Radiosity and Global Illumination*, Morgan Kaufmann Publishers, Inc., 1994.

[SNY 93] SNYDER J.P. *Flattening the Earth – Two Thousand Years of Map Projections*, The University of Chicago Press, Chicago, 1993.

[SNY 01] SNYDER J.M., MITCHELL D., Sampling efficient mapping of spherical images, Microsoft Technical Report, submission *to SIGGRAPH 2001*: papers_0250, 2001.

[STE 61] STEVENS S.S., "To honor fechner and repeal his law", *Science, New Series*, vol. 133, no. 3446, pp. 80–86, 13 January 1961.

[SUM 92] SUMPNER W.E., "The diffusion of Light", *Proceedings of the Physical Society of London*, vol. 12, no. 10, 1892.

[SWI 06] SWINBANK R., PURSER R.J.,, "Fibonacci grids: a novel approach to global modelling", *Quarterly Journal of the Royal Meteorological Society*, vol. 132, pp. 1769–1793, 2006.

[TAY 95] TAYLOR R.P., LUCK R., "A case study of view factor rectification procedures for diffuse-gray radiation enclosure computations", *6th Annual Thermal and Fluids Analysis Workshop*, Cleveland, OH, 1993.

[TRE 87] TREGENZA P.R., "Subdivision of the sky hemisphere for luminance measurements sky", *Ligthing Research and Technology*, vol. 19, pp. 13–14, 1987.

[VER 75] VERSPRILLE K.J., "Computer-aided design applications of the rational B-spline approximation form", Electrical Engineering and Computer Science – Dissertations, Paper 262, 1975. Available at http://surface.syr.edu/eecs_etd/262.

[VOL 10] VOLLMER M., MÖLLMANN K.-P., *Infrared Thermal Imaging: Fundamentals, Research and Applications*, Wiley-VCH, pp. 612, 2010.

[VUE 06] VUEGHS P.I., BECKERS P., "Presentation of the hemisphere method", *Advanced Computational Methods in Heat Transfer IX, Transaction: Engineering Science*, vol. 53, pp. 121–130, 2006.

[WHI 98] WHITE D., KIMERLING A.J., SAHR K., *et al.*, "Comparing area and shape distortion on polyhedral-based partitions of the sphere", *International Journal of Geographical Information Science*, vol. 12, no. 8, pp. 805–827, 1998.

[WHI 80] WHITTED T., "An improved illumination model for shaded display", *Communications of the ACM (Association for Computing Machinery)*, vol. 23, no. 6, pp. 343–349, June 1980.

[WHY 52] WHYTE L.L., "Unique arrangements of points on a sphere", *The American Mathematical Monthly*, vol. 59, no. 9, pp. 606–611, November 1952.

[WOO 94] WOOLDRIDGE C., HAYMAN S., "Cloud cover assessment grids for whole-sky images", *Lighting Research & Technology*, vol. 26, no. 1, pp. 9–12, 1994.

[XEN 71] XENAKIS I., *Musique. Architecture.*, Casterman, Tournai, Belgium, 1971.

[ZHA 91] ZHANG W.H. Sensitivity analysis and structural shape optimization by finite element method, PhD Thesis, University of Liège, Belgium, 1991.

[ZIE 71] ZIENKIEWICZ O., *The Finite Element Method in Engineering Science*, McGraw-Hill, 1971.

Index

N, O, P, R

S, T, V, W, Z